The World's Pharmaceutical Industries

The World's Pharmaceutical Industries

An International Perspective on
Innovation, Competition and Policy

Prepared for the
United Nations Industrial Development Organization

Robert Ballance, János Pogány and Helmut Forstner

HD
9665.5
.B35
1992

Edward Elgar

Published by
Edward Elgar Publishing Limited
Gower House
Croft Road
Aldershot
Hants GU11 3HR
England

Edward Elgar Publishing Company
Old Post Road
Brookfield
Vermont 05036
USA

A CIP catalogue record for this book
is available from the British Library

Library of Congress Cataloging-in-Publication Data
Ballance, Robert.
　　The world's pharmaceutical industries: an international perspective on inno-
vation, competition, and policy/by Robert Ballance, János Pogány, and Helmut
Forstner·
　　　p. cm.
　　"Prepared for the United Nations Industrial Development Organization".
　　Includes bibliographical references.
　　1. Pharmaceutical industry.　2. Pharmaceutical industry—
Technological innovations.　3. Pharmaceutical industry—Government
policy.　I. Pogány, János.　II. Forstner, Helmut.　III. United
Nations Industrial Development Organization.　IV. Title.
　　[DNLM: 1. Drug Industry—economics.　QV 736 B188w]
　　HD9665.5.B35　1992　　　338.4'76159—dc20
　　DNLM/DLC　　　　　　　　　　　　　　　　　　　　　91–42995
　　for Library of Congress　　　　　　　　　　　　　　　　　　CIP

ISBN 1 85278 646 9

Printed in Great Britain by
Billing & Sons Ltd, Worcester

Contents

Tables

Figures

Boxes

Abbreviations

AIDS	acquired immune deficiency syndrome
DNA	deoxyribonucleic acids
DPCO	Drug Prices Control Order
EC	European Community
ECMA	European Community Market Authorization
EDL	essential drug list (Nigeria)
EFTA	European Free Trade Association
EMCA	European Medicines Control Agency
EPI	Extended Programme on Immunization (WHO)
FDA	Food and Drug Administration (United States)
FDI	foreign direct investment
GDP	gross domestic product
GLP	good laboratory practices
GMP	good manufacturing practices
GNP	gross national product
HAI	Health Action International
HMO	health maintenance organization
ICI	Imperial Chemical Industry Limited (United Kingdom)
IDA	International Dispensary Association
ILO	International Labour Organization
IMF	International Monetary Fund
IMS	International Medical Services
IND	Investigational New Drug Application (United States)
INN	international non-proprietary name
IOCU	International Organization of Consumer Unions
J&J	Johnson and Johnson
JIT	just-in-time inventory control
JPMA	Japan Pharmaceutical Manufacturers Association
MHW	Ministry of Health and Welfare (Japan)
MITI	Ministry of International Trade and Industry (Japan)
MVA	manufacturing value added

NCE	new chemical entity
NDA	new drug application (United States)
NHS	National Health Service (United Kingdom)
NME	new molecular entity
OECD	Organization for Economic Co-operation and Development
OTC	over-the-counter drug
P&G	Proctor and Gamble
PIC	pharmaceutical inspection convention
PMA	Pharmaceutical Manufacturers' Association (United States)
PMAG	Pharmaceutical Manufacturers' Association of the Federal Republic of Germany
PPMA	Pakistani Pharmaceutical Manufacturers' Association
PRS	Price Regulation Scheme (United Kingdom)
PTA	Preferential Trade Area
SITC	standard international trade classification
SKB	SmithKline Beecham
SSCI	Swiss Society of Chemical Industries
UNCTAD	United Nations Conference on Trade and Development
UNCTC	United Nations Centre on Transnational Corporations
UNDP	United Nations Development Programme
UNICEF	United Nations Children's Fund
UNIDO	United Nations Industrial Development Organization
WHO	World Health Organization
WIPO	World Intellectual Property Organization

Explanatory Notes

The following classification of economic groupings is used in the text, and in most tables, in conformity with that used by the Statistical Office of the United Nations Secretariat: 'Developing countries' includes all countries, territories, cities and areas in Africa (except South Africa), Latin America, East Asia (except Japan), South Asia and West Asia (except Israel). 'Developed market economies' includes Northern America (Canada and the United States of America), Europe (other than Eastern Europe), Australia, Israel, Japan, New Zealand and South Africa. 'Centrally planned economies' includes Bulgaria, Czechoslovakia, the German Democratic Republic, Hungary, Poland, Romania and the Union of Soviet Socialist Republics. (For purely statistical reasons, Yugoslavia is listed in several tables as a domestic market economy.) Unless otherwise specified, 'world' excludes Albania, China, the Democratic People's Republic of Korea, Mongolia and Viet Nam. In some tables the classification may differ slightly from the above, depending on the source cited.

Unless arranged otherwise for statistical reasons, countries are generally listed in alphabetical order. In listings and tables, inclusion or exclusion of a particular country may have been dictated by considerations of the availability of comparable data; it does not necessarily express a judgement concerning the stage reached by the country in the development process.

The Federal Republic of Germany, which is cited very frequently in the present publication, is referred to as 'Germany, Federal Republic of' in listings and tables (United Nations usage). To avoid unnecessary awkwardness, however, this form is used only when five or more countries are listed together.

Unless otherwise indicated, 'manufacturing' includes the industry groups listed under Major Division 3 in *Indexes to the International Standard Industrial Classification of All Economic Activities* (United Nations publication, Sales No. E.71.XVII.8).

Mention of commercial enterprises does not imply endorsement of those enterprises by the United Nations.

International Standard Industrial Classification of all Economic Activities (ISIC) code numbers are accompanied by a descriptive title (for example, ISIC 352: 'Manufacture of other chemical products'). For considerations of space, however, the description is sometimes shortened (for example, ISIC 352 may be described simply as 'Other chemical products').

Dates divided by an oblique (1970/71) indicate a crop year or a financial year. Dates divided by a hyphen (1970–5) indicate the full period involved, including the beginning and end years. References to dollars ($) are to United States dollars, unless otherwise stated. References to tons are to metric tons, unless otherwise specified.

Annual rates of growth or change are based on data for each year throughout the period indicated and are calculated using a semilogarithmic regression over time, unless otherwise specified.

In tables (i) apparent arithmetical discrepancies, such as percentages that do not add precisely to totals, result from rounding of basic data or from differences in rounding of figures known to different degrees of precision; (ii) three points (...) indicate that data are not available or are not separately reported; (iii) a rule (–) indicates that the amount is nil or negligible; (iv) a blank indicates that the item is not applicable; (v) a minus sign (-) before a figure denotes a deficit or decrease, unless otherwise indicated.

Acknowledgements

This book draws upon a wide range of material of a technical, empirical and economic nature. Accordingly the contributors include a number of individuals in addition to the principal authors. The contributions of several individuals from outside UNIDO are gratefully acknowledged, though none bears responsibility for any remaining errors. Dr Bengt Jönsson of Linköping University and Mr Stuart Sinclair of Temple, Baker and Sloane both served as consultants, supplying papers which were adapted by the authors and incorporated into the book. Dr Sanjaya Lall (Oxford Institute of Economics and Statistics) served as an external adviser and provided valuable criticisms and comments. Finally the contribution of statisticians and other members of the Industrial Statistics and Sectoral Surveys Branch of UNIDO was substantial and greatly appreciated.

This publication has been prepared by Robert Ballance, János Pogány and Helmut Forstner, who are responsible for the views and opinions expressed therein. These views and opinions do not necessarily represent those of other persons or institutions mentioned here.

1. The World's Pharmaceutical Industries: A Global Map

The foundations of the modern pharmaceutical industry date back to the development of orthodox methods of drug research shortly after the Second World War. Although the industry's history is comparatively brief, its accomplishments are impressive. A flood of life-saving drugs has emerged from the world's laboratories over the past four decades. By combating many fatal diseases and eradicating others, drug producers have helped to alter mortality patterns in many parts of the world.

These achievements alone would be sufficient to make the industry a worthy subject for study. However there are a number of additional reasons why it merits a detailed investigation. One is the highly politicized environment in which drug companies operate. Their performance, whether measured in terms of product development, prices, safety or efficacy, is a vital determinant of health. Regulations vary from country to country, but all governments intervene extensively to ensure that national standards are met.

The attitudes of consumers and regulators hardened when several drugs were belatedly found to have disastrous side-effects. The industry's proponents and critics now tend to operate in a highly-charged atmosphere where questions of policy and practice are depicted in narrow and stark terms. Pharmaceuticals, however, embodies a complex and highly interrelated set of activities. Many issues can only be addressed in the rather broad context afforded by a study of this type.

A second noteworthy feature is the industry's unique configuration. The development of powerful new drugs led to an explosion in demand, demonstrating the crucial role to be played by research and innovation. Producers were just as quick to recognize that the market value of new products could be protected by patent and promoted through the use of brand names. This combination of research prowess and marketing power helps to explain the industry's internal structure.

1

At one end are large teams of organic chemists, biochemists, biophysicists and pharmacologists who use all the tools of modern science in their search for new drugs. At the other are huge – and highly competitive – networks for distribution and promotion which operate in markets where performance is closely monitored by governments. Sandwiched between these two crucial parts of the industry is the production of drugs.

The industry's unique blend of scientific knowledge, manufacturing skills and marketing tactics means that most issues must be seen from more than one perspective. Accordingly the discussion in this book spans a number of disciplines and its contributors are drawn from several different fields. They include specialists in pharmaceuticals and chemistry, economists with a particular interest in industrial organization, international trade and health, and business analysts concerned with matters of corporate strategy.

A third noteworthy characteristic is the extent of internationalization and diversity which exists. A great deal of the industry's research and production occurs in only a very few countries. However the markets for drugs are global and companies have begun to scatter their operations around the world as they have grown. The types of pharmaceutical firms are just as varied and diverse as their markets. A small number of companies dominate every phase but the industry's membership is actually very large. In addition to the multinationals and their subsidiaries, it includes many small and medium-sized firms, niche producers and other specialists identified by their research and marketing strengths or by the specific types of drugs they produce. The extent of internationalization and firm specialization is growing, although most issues tend to be discussed in terms of only a few countries or a particular subset of firms (usually the multinationals). The priorities of this book are somewhat different: special attention is given to conditions in developing countries, while the role of small and medium-sized firms is examined along with that of the multinational. The reasons for this orientation are obvious, given that an international organization has provided the resources and support for the study. It is also justifiable in view of the industry's significance for consumers and governments in developing countries.

One final point about the book should be made before we conclude this introduction. Most drug companies are extremely secretive and the markets they serve are highly fragmented. Because of these characteristics, the amount of information which is available on the industry is limited. A high priority has been placed on the compilation of data relating to production, trade, investment, costs, research spending and other important activities. The intention is not to produce a snapshot of the industry at a recent point in time. Such information would be of limited use since the industry is at present in the midst of a major transformation. Instead, the purpose is to gain some impression of long-term changes in key parts of the industry which can then serve as some guide for analysis of issues and prospects.

These priorities mean that a great deal of time has been spent in gathering empirical information. One of the main sources has been the UNIDO data base. The data base has been constructed from the annual responses of national statistical offices in more than 150 countries. Additional information has been collected from industry-specific literature, national producer associations and studies carried out by government agencies and international institutions. Company reports served as a third source of information and a special data base was developed in order to utilize these statistics. Finally UNIDO operates a large programme of technical assistance for the pharmaceutical industry in developing countries and is active in promoting links between these companies and others in industrialized countries. Information collected in the course of this work has also been used.

The following section sets up a framework for the discussion in later chapters. A broad outline of the pharmaceutical industry is presented and some indication of the study's priorities is given.

A TYPOLOGY OF PHARMACEUTICAL INDUSTRIES AND PRODUCTS

Several of the world's major pharmaceutical companies can trace their lineage back to the nineteenth century, although the industry itself only began to assume importance after the Second World War. Its growth over the past few decades has been remarkable. World production (measured in 1980 dollars) has increased more than twofold since 1975 and in 1990 stood at $150 billion. Roughly 60 countries now produce at least $100 million worth of pharmaceuticals each year. The markets for drugs have grown almost as rapidly. On a per capita basis, world consumption rose from $17 in 1975 to $29 in 1990.

Such figures are impressive but they do not provide an accurate picture of the industry's geographical configuration. In fact the bulk of the world's pharmaceuticals are manufactured in a very few industrialized countries. The pattern of drug consumption is much the same: over three-quarters of all medicines are sold in industrialized countries, with the remainder being purchased by households in developing countries. Pharmaceuticals has clearly begun to assume an international character but the industry is still not a global one in the same sense as textiles, food processing, clothing or even steel.

The contrasts are equally great when attention turns from countries to firms. A small number of about 50 multinationals account for two-thirds of the world's production and exports each year. The largest of these is Merck. The company's revenues in 1990 were $7.7 billion, an amount exceeding the entire pharmaceutical production of Latin America in that year. Alongside these giants are thousands of small and medium-sized drug firms. More than two-fifths of the companies operating in industrialized countries have annual sales of less than $25 million and the proportion is much higher in developing countries.

The variety of products which the industry produces is another important feature. About 20 000 different medicines are sold in huge markets like the United States or Japan. More than 10 000 products are available in the bigger developing countries – for example, Brazil, Mexico or the Republic of Korea – and the number is almost as great in many smaller industrialized and developing countries. It follows that the degree of product differentiation is great, depending not only on product characteristics but on methods of distribution, aspects of national policy and the effectiveness of promotional campaigns.

Not surprisingly several products are available in each country to treat a particular ailment. Not all are of equal importance, however. A single product may account for as much as a quarter of total sales and the five largest suppliers may claim two-thirds of the domestic market or even more. For certain diseases the same products dominate the market throughout the world, but in other cases the leadership changes from one country to the next.

The foregoing are only a few of the many characteristics which attest to the industry's diversity. The list is sufficient, however, to suggest that generalizations about various aspects will be applicable only to specific parts of the industry. Such heterogeneity is particularly important when issues of public policy, corporate strategy or cost structure are considered. In order to facilitate the discussion in following chapters, a country typology has been developed.

The country typology singles out three types of drug producers: large, integrated corporations, innovative companies and reproductive firms. The integrated corporations are multinationals engaged in all three stages of drug production – research, manufacture and distribution. These firms are distinguishable in several ways. First, they are exceptionally large; annual sales are at least $200 million and in many cases exceed $1 billion. Second, integrated firms place a particularly high priority on product development, generating the new molecular entities (NMEs) which are essential for their research and production activities. Third, they adhere to several well-defined methods of operation. Patents are secured for inventions at global level, medicinal chemicals are distributed through subsidiaries and licensees, intermediate inputs are purchased only from approved vendors, and pharmaceutical preparations are generally sold under brand names on the private market.

Innovative companies are easily distinguished from integrated companies. They are capable of discovering and developing NMEs but typically produce patent-expired drugs. Annual sales are modest, ranging between $25 and $200 million. Revenues of this magnitude are not sufficient to fund the massive research programmes and distribution systems that the vertically integrated corporation operates. Instead the innovative firm that develops an NME will usually resort to licensing arrangements to market the product abroad.

The lack of significant research or distribution facilities does not mean that the innovative firm is excluded from international markets. Some operate a few foreign subsidiaries, while others maintain scientific and trade offices in overseas markets and participate in joint ventures with foreign partners. Many are significant exporters, either selling drugs directly through their own channels or through international trading houses on the open market.

Reproductive firms complete this characterization of pharmaceutical companies. These are either small, family-owned enterprises or publicly-owned companies of medium size. Lacking any in-house research capacity, they utilize the scientific and technological knowledge developed by others to manufacture their product. It follows that the drugs they produce are not protected by patent. The operations of reproductive firms will differ in several ways, however. Special packing materials and some of the inputs (for example, active ingredients) are purchased either through international tenders or from approved suppliers while the medicines themselves are sold under brand names or as cheaply-priced generics. The sales efforts of reproductive firms are divided between the private and public market, although the latter is frequently the more important, particularly in developing countries.

The characterization of firms described here is not exhaustive (see Box 1.1) but it is sufficient to yield a reasonably accurate typology of the world's pharmaceutical industries. Four broad groups of national producers can be identified according to the types of firms that exist in each country. Table 1.1 indicates the membership of these four groups while Table 1.2 gives some general indicators for production, exports and consumption.

The countries with the largest and most sophisticated pharmaceutical industries are those in category A. The prominence of this group rests on two cornerstones. First, all types of firms exist in each of these countries but it is the large, integrated corporations which dominate in every case. The leadership of these companies is the main reason for the group's overwhelming contribution to world production and exports. The share of exports has declined but this does not mean that the competitive impetus has shifted to other parts of the world. Most of the integrated producers are multinationals with large foreign investments in other countries which are members of the same group. Exports, therefore, have not kept pace with production because foreign subsidiaries have taken over some of these markets.

The second factor explaining the industry's heavy concentration in these countries is the fact that their own drug markets have grown exceptionally fast. Levels of per capita consumption increased dramatically between 1975 and 1990 and are many times greater than those in other parts of the world. Several reasons for this are explored in later chapters, but two of the more obvious factors can be mentioned here. One is the fact that the populations of these countries are aging rapidly. As this occurs, demand shifts towards more expensive types of medicines and the frequency of drug consumption rises. The other is that the governments of these countries established very generous systems of public health care when national incomes were rising and the proportion of elderly was comparatively small. The public

---

Box 1.1   Small, research-based drug firms: a unique group

One group of firms which is not considered here consists of small and medium-sized companies engaged in research-intensive activities. They are founded in order to exploit a single or a small number of patents for the development of a unique drug.

The most important of these firms are engaged in genetic engineering. The drugs they produce are not really NMEs since they are generally identical, or very similar, to substances which occur naturally in the body. The best-known of these firms are Biogen, Biotech, Cetus and Genetech. All were founded at an early stage of product development as joint ventures between academic centres, venture capital entrepreneurs and/or integrated pharmaceutical firms. Such firms typically incur operational losses at early stages of operation since their research expenditures exceed revenues earned from royalties (see Box table 1.1). Modern biotechnology firms may launch patented products at home and conclude licensing arrangements for international marketing. Most become targets for acquisition once their particular product is established in major markets. As a result many are now becoming more closely integrated with mainstream firms in the pharmaceutical industry.

Another, much smaller subset consists of firms such as Alza, Elan and KV Pharmaceuticals which specialize in the development of new therapeutic systems providing controlled-release oral and transdermal pharmaceutical preparations for new and established drugs. Still other firms offer products and services to the pharmaceutical industry on a contractual basis. These firms may specialize in synthesis of new compounds, *in vivo* studies, product registrations and so on.

Box table 1.1   Characteristics of small and medium-scale pharmaceutical firms (US$ millions)

Company (headquartered)	Year	Operating revenues $	Profit $	Profit %	R&D spending $	R&D spending %
Alza (USA)	1988	84.2	17.0	20.2	29.4	34.9
Biogen (USA)	1988	30.2	1.2	3.9	24.2	80.0
Chiron (USA)	1988	21.7	11.2	51.8	28.1	129.9
Genetech (USA)	1988	334.8	20.6	6.1	132.7	39.6
Genetics (USA)	1988	29.7	-11.2	-37.9	47.6	160.2
Elan (IRE)	1988	10.6	-3.4	-32.5	2.2	21.2
Liposome Technology (USA)	1988	4.9	-3.3	-67.1	6.3	128.1
Liposome Company (USA)	1987	4.0	-6.3	-155.1	8.0	199.1
Nova (USA)	1987	14.3	-4.4	-31.0	13.2	92.0
Pharmatec (USA)	1987	0.8	-1.9	-251.7	1.9	253.5
Praxis (USA)	1987	13.0	0.3	2.1	9.8	75.0
Average		49.8	1.8	3.6	27.6	55.3

Source: *Pharmaprojects* (1989).

Table 1.1   A typology of the world's pharmaceutical industries

---

A   Countries with a sophisticated pharmaceutical industry and a significant research base

Belgium	Netherlands
France	Sweden
Germany	Switzerland
Italy	United Kingdom
Japan	United States

B   Countries with innovative capabilities[1]

Argentina	Ireland
Australia	Israel
Austria	Mexico
Canada	Portugal
China	Republic of Korea
Denmark	Spain
Finland	USSR
Hungary	Yugoslavia
India	

C   Countries/areas with reproductive capabilities

C1   Those producing both therapeutic ingredients and finished products

Bahamas[2]	Indonesia[3]
Bolivia	Macau[4]
Brazil	Norway
Bulgaria	Poland
Cuba	Puerto Rico
Czechoslovakia	Romania
Egypt	Turkey

C2   Those producing only finished products

Afghanistan	Democratic People's	Iraq
Albania	Republic of Korea	Jamaica
Algeria	Dominican Republic	Jordan
Angola	Ecuador	Kenya
Bangladesh	El Salvador	Kiribati
Barbados	Ethiopia	Kuwait
Belize	Fiji	Lebanon
Benin	Gambia	Lesotho
Brunei	Ghana	Liberia
Cambodia	Greece	Madagascar
Cameroon	Guatemala	Malawi
Cape Verde	Guyana	Malaysia
Chile	Haiti	Mali
Colombia	Honduras	Malta
Costa Rica	Hong Kong	Mauritius
Côte d'Ivoire	Iran (Islamic	Mongolia
Cyprus	Republic of)	Morocco

C2   continued

Mozambique	Saudi Arabia	Tunisia
Myanmar	Seychelles	Uganda
Namibia	Sierra Leone	United Arab Emirates
Nepal	Singapore	United Republic of
New Zealand	Solomon Islands	Tanzania
Nicaragua	Somalia	Uruguay
Niger	South Africa	Venezuela
Nigeria	Sri Lanka	Viet Nam
Pakistan	Sudan	Yemen
Panama	Syrian Arab Republic	Zaire
Papua New Guinea	Taiwan Province	Zambia
Paraguay	Thailand	Zanzibar
Peru	Tonga	Zimbabwe
Philippines	Trinidad and Tobago	

D   Countries/areas without a pharmaceutical industry

Andorra	French Guyana	New Caledonia
Antigua and Barbuda	French Polynesia	Niue
Aruba	Gabon	Oman
Bahrain	Greenland	Qatar
Bermuda	Grenada	Reunion
Bhutan	Guadeloupe	Rwanda
Botswana	Guam	St Kitts and Nevis
British Virgin	Guinea	St Lucia
Islands	Guinea-Bissau	St Vincent-
Burkina Faso	Iceland	Grenadines
Burundi	Laos	Samoa
Central African	Libyan Arab Jamah.	San Marino
Republic	Liechtenstein	Sao Tome and Principe
Chad	Luxembourg	Senegal
Comoros	Maldives	Suriname
Congo	Martinique	Swaziland
Cook Islands	Mauretania	Togo
Djibouti	Mayotte	Tuvalu
Dominica	Micronesia	US Virgin Island
Equatorial Guinea	Nauru	Vanuatu
Faeroe Islands	Netherland Antilles	Western Samoa

Notes

[1] Each country in this group discovered and marketed at least one NME between 1961 and 1990.

[2] Firms produce therapeutic ingredients (medicinal chemicals) for export.

[3] Every foreign-owned factory is required to produce at least one therapeutic ingredient within five years of start-up.

[4] Hovione's (Portugal) subsidiary produces antibiotics and corticosteroids for export.

Source: UNIDO.

sector at present accounts for more than half of all drug expenditures in all of the countries in Group A except the United States.

The world's leading producers dominate every phase of the industry but nowhere is their leadership more pronounced than in the field of research. Unlike the production and distribution of drugs, research centres have not migrated to other parts of the world. This phase of the industry is highly centralized, being located either in the country where the firm is headquartered or in one of the other industry leaders. Several of the governments concerned are active supporters of research, sometimes providing funding which exceeds the sums spent by the country's multinationals. The combination of public and private funding has borne impressive results. More than 90 per cent of the new drugs which the industry has marketed since 1960 were discovered and developed in one of the countries in category A.

Among the countries in group B the industry consists only of innovative and reproductive firms. The fact that large, integrated firms do not exist in this group means that research, production and distribution capabilities are modest in comparison with the industry leaders. Innovative companies have begun to compete more aggressively in export markets, however, and it is here that the group has its greatest impact. Several members of the group – Austria, China, Denmark, Hungary, Spain and Yugoslavia – are now among the world's 20 largest exporters, though only China boasts levels of production approaching those in France, Germany or the United Kingdom.

The pharmaceutical industries in group C are populated solely by reproductive firms. They may be locally-owned firms, subsidiaries of multinationals or joint ventures between indigenous and foreign companies. The countries in group C are not originators of new drugs and, if they were, there would be no overseas facilities for distribution.[1] In several of these countries the firms do no more than produce the finished product from imported inputs. The extent of foreign ownership varies, depending on the country's policies, the size of the market and the strategies of the multinationals. In some instances foreign subsidiaries claim no more than 20 per cent of the domestic market, a figure which is not much different from that in many industrialized countries. In other countries, they account for a much larger proportion of that market – occasionally more than three-quarters.

Very little can be said about the last group of countries in this typology. Production is nil and the domestic market, which is very

Table 1.2   *The global pattern of pharmaceutical production, exports and consumption, 1975 and 1990*[a]

Country group (no. of countries)	Share in world production (%)		Share in world exports (%)		Per capita consumption (in 1980 $)	
	1975	1990	1975	1990	1975	1990
A Countries with a sophisticated pharmaceutical industry and a significant research base (10)	60	69	78	68	65.8	150.5
B Countries with an innovative pharmaceutical industry (16)	28	22	19	27	9.6	12.2
C Countries/areas with reproductive capabilities (91)	12	9	4	5	7.4	9.4
D Countries/areas without a pharmaceutical industry (43)	–	–	–	–	5.4	8.2

[a] All figures are based on data in 1980 dollars and refer to pharmaceutical preparations.

Source: Table 1.1 and UNIDO, based on data reported by national pharmaceutical manufacturers' associations; IMS; United Nations.

small, is serviced entirely through imports. The average level of per capita consumption is comparable to that in group C but the amounts of medicines consumed in some countries are only half this figure.

One last point remains to be noted. The typology described here does not explicitly take into account the multitude of products nor the degree of product differentiation which exists. The terminology and product nomenclatures used within the pharmaceutical industry are overlapping and frequently confusing to all but the industry specialist. Much of this terminology can be avoided, though certain product arrangements figure prominently in the book, and it is helpful if the reader is aware of these from the outset.

Table 1.3 sets out four types of product arrangements which re-appear in later chapters. The first of these is indicative of the forms of competition which exist within the industry. The development of a new drug begins with the identification of an NME, which is also referred to as a new chemical entity (NCE). These compounds are usually patented as soon as they are discovered – even before their usefulness, if any, has been determined. Over 95 per cent of NMEs do not survive the subsequent battery of tests designed to determine safety and efficacy but those that do eventually lead to a new, or improved, drug.

Large, integrated firms must have a steady stream of new or improved drugs to justify their huge investments in research and distribution. It follows that countries with a highly advanced and fully integrated pharmaceutical industry (that is, group A) are the main source of new drugs. Between 1960 and 1988 roughly 2000 NMEs were successfully marketed as new drugs; over 90 per cent were produced by integrated corporations or innovative companies based in one of these ten countries. The remainder were developed by innovative firms operating in one of the countries in group B.

In the case of multiple-source products, price rather than product efficacy or originality is the basis for competition. All these drugs are referred to here as generics. No internationally accepted definition of the category exists, but for purposes of this book generics are regarded as drugs which are no longer protected by patent or are non-patentable. The global market for generic drugs in 1991 was $15 billion, or roughly 8 per cent of world sales.

Two types of generic products are noted in Table 1.3, depending on whether a non-proprietary name or a brand name is used. National

Table 1.3 Alternative groupings of pharmaceutical products

Form of competition	End-use	Choice-maker and distribution channel	Type of market exclusivity
1. Single-source drugs: Innovative medicines produced from NMEs or NCEs	**1. Medicinal chemicals:** Intermediate products produced by synthesis, fermentation or extraction from natural resources	**1. Prescription drugs:** supplied by retail pharmacies, hospitals (publicly or privately-owned), physicians and other outlets	**1. Patent-protected:** includes NMEs, innovative extensions and patentable chemical variations as well as patents on technological processes
2. Multiple-source drugs: all of these products are regarded as generics. Some are sold under a non-proprietary name. Others are sold under a brand name in order to distinguish the product from therapeutically equivalent versions distributed by competitors	**2. Pharmaceutical preparations:** finished products such as tablets, injections, solutions and capsules	**2. Over-the-counter drugs:** products vary depending on national regulations; they may be distributed through several different professional and commercial outlets	**2. Other policy measures:** temporary monopolies for new drugs; ownership control
	3. Diagnostics: a heterogenous group of finished products derived from chemical or biological sources	**3. Parapharmaceutical products:** homeopathic versions may be obtained through pharmacies; special outlets are sometimes available for traditional medicines produced in factories	**3. Firm-specific measures:** control of key medicinal chemicals; development of sophisticated manufacturing knowhow
	4. Biologicals: mainly consist of vaccines such as those used for the prevention of childhood diseases		

Source: UNIDO

13

and international bodies assign non-proprietary names to each drug and a manufacturer is free to select one of these to identify its product. Thus it is possible that several manufacturers will be producing a drug that is identical both in name and therapeutic qualities. Accordingly competition is purely in terms of price and product availability. Other firms competing in the same market can choose to differentiate their version of the product from others by using a brand name which must be registered with a government office.[2]

Pharmaceutical products may also be described according to their end-use. Four broad classes of drugs are identified in this manner although only two – pharmaceutical preparations and medicinal chemicals – are considered in this publication (see Box 1.2). The most common forms of pharmaceutical preparations are tablets, injections, solutions and capsules. These are finished products which are known to all consumers and are produced by integrated, innovative or reproductive firms alike.

Medicinal chemicals are a more important but less familiar part of the industry. They are the active ingredients used to produce the drug. Many medicinal chemicals are widely available since they are produced from organic chemicals which are purchased on open markets and well-known technologies are used in the production processes. Known as commodity medicinal chemicals, these inputs are manufactured by all three types of firms considered here.

Medicinal chemicals other than the commodity versions are not so common. Some are NMEs which are patent-protected. Others are not distributed because they are critical to a firm's profitability or because their manufacture requires sophisticated methods of synthesis, fermentation or extraction. Companies with access to and mastery of the key technologies often regard these chemicals as strategic inputs since their availability accords the producer some advantage over competitors. Both large, integrated companies and innovative firms prefer to manufacture these inputs in-house to ensure that they have a secure 'captive' supply. Innovative firms are not always in a position to adopt this approach, however. They may lack the necessary expertise, the costs of environmental protection can be too high or the quantities required may be insufficient for the firm to produce economically.

The third product arrangements in Table 1.3 are recognized differences in distributional methods. When pharmaceuticals are pictured in terms of the distribution channels and choice-makers (that is, the

---

Box 1.2   *Diagnostic and biological pharmaceuticals*

These two types of drugs are not considered in this publication. Diagnostic products consist mainly of kits to determine fertility, blood glucose, gastrointestinal bleeding and so on. The markets for these products are growing as consumers have become more sophisticated, but use is still confined to a small number of industrialized countries. In the case of biologicals, vaccines are the dominant product group. Their markets are small in comparison to other drugs, although vaccines are an important part of public health programmes, particularly in developing countries. Most vaccines are used in the prevention of childhood diseases – diphtheria, measles, pertussis, poliomyelitis, tetanus and tuberculosis – but more and more elderly people are becoming aware of the benefits of being vaccinated against influenza and pneumonia. Niche markets include vaccination against yellow fever, cholera, hepatitis and meningitis transmitted by ticks. The development of an effective and safe vaccine for the prevention of acquired immune deficiency syndrome (AIDS) would lead to an exponential growth in demand.

---

doctors or health professionals who prescribe the drug), three classes of products can be distinguished. Prescription drugs and over-the-counter drugs (OTCs) play a prominent role in this book, though parapharmaceutical products are a minor part of the industry and not considered (see Box 1.3).

Prescription drugs account for the bulk of the medicines sold. The actual products which fall into this category are determined by national health officials and the purchaser may or may not be reimbursed through the public health care programme. In many industrialized countries prescription drugs are available only through retail pharmacies or hospitals. Physicians are also part of the distribution system in Japan and in most developing countries. Primary health care workers are another important outlet in developing countries. However regulations governing distribution are not strictly enforced in many developing countries and prescription drugs can be purchased easily from street vendors.

OTCs have a small share of the market but their importance is growing. Because OTCs are for self-treatment of minor ailments, they must be safe to use without a doctor's advice and supervision. These are multiple-source drugs which are produced by all three categories of firms. Price controls are lenient, or else the supplier is free to set its own price. This relaxed attitude results from the fact that the costs of

The World's Pharmaceutical Industries

Box 1.3 Parapharmaceutical products

This product class consists of homoeopathic products and factory-made traditional medicines. Homoeopathic products can be registered for sale if they are safe but scientific evidence of effectiveness is not required. None of the larger pharmaceutical companies are suppliers of these products. Traditional medicines, however, are important in several Asian markets (see Box table 1.3). Evidence with regard to safety and experience is based on thousands of years of experience. Sellers in industrialized countries must provide scientific proof of safety and stable quality but evidence of efficacy is seldom required.

Box table 1.3   Sales of modern pharmaceuticals and factory-made traditional medicines in China, 1975–90 (US$ millions and per cent of total sales)

Year	Modern pharmaceuticals		Traditional medicines	
	$	%	$	%
1975	2 845	89.1	350	10.9
1980	4 515	86.6	696	13.4
1985	3 368	82.0	739	18.0
1990	4 092	80.4	998	19.6

Source: UNIDO data base.

OTCs are not reimbursable and therefore do not affect public health care expenditures. Finally mass media advertising is common and several distribution channels (including pharmacies) are used.

The last of the product arrangements considered here refers to the degree of market exclusivity which results from policy decisions or proprietary expertise. Patent protection is distinguished from other types of policy because it has been a major international issue in the pharmaceutical industry for several decades. Innovative chemical extensions, which are therapeutically improved versions of the original NME, can also be patented. It is also possible to patent other chemical variations. These, however, may have therapeutic qualities which differ very little from the original product and so are referred to as 'me too' drugs. Finally patents are granted for certain process technologies and can offer the originator as much protection from competition as a product patent.

There are a number of other government policies that alter the degree of market exclusivity. Those noted in Table 1.3 are only two examples. The most common form of market exclusivity is ownership control. Many firms, particularly those in eastern Europe and the USSR but also in developing countries, are state-owned. They frequently have a monopoly position although prices are usually strictly controlled and kept at artificially low levels.

Some governments also attempt to correct for market failure by granting exclusivity for a limited period of time. They may do this even if the chemical compound is already known and is not patentable. Known as 'orphan drugs', these medicines are not marketed because the afflictions they treat are too rare for pharmaceutical companies to bother with. The United States government, for example, encourages companies to develop orphan drugs by granting seven years of market exclusivity, irrespective of the patent situation. The law has had its successes but has also created other types of problems (see Box 1.4).

Multinationals and innovative firms may attain a degree of market exclusivity not through policy but by dint of superior in-house expertise. Pharmaceutical companies specializing in production processes using toxic or dangerous chemicals or involving high-yield fermentation methods frequently benefit in this way. The firms which produce these medicinal chemicals treat them as captive inputs rather than selling them on the open market. Such practices may persist long after the relevant product patents have expired since it is clear that competitors would have great difficulty in replicating the production process in any case.

In conclusion, this brief excursion through the pharmaceutical industry serves as a point of departure for the study. The book proceeds as follows. Chapter 2 builds on the framework presented here by providing a survey of global trends in drug production and consumption. The survey leaves little doubt that the industry is currently passing through a period of turmoil and change. As these changes occur the competitive position of different sets of producers may be altered. Chapter 3 addresses this subject through an analysis of export performance, innovative leadership and foreign investment. Both the industry and its consumers rely heavily on the continued development of new drugs. Research and product development are accorded a particularly high priority in the case of pharmaceuticals and this topic

---

Box 1.4   Promoting the development of orphan drugs: successes and abuses

In 1983 the United States Congress passed a law granting tax breaks and a seven-year monopoly to companies producing orphan drugs. The drugs were defined as having annual sales of less than $5 million and a potential market of fewer than 200 000 patients. The law has had its successes, such as a treatment for porphyria – a painful and debilitating disease that afflicts only 100 people. It has also proved to have shortcomings. Some companies make use of their monopoly position to extract exorbitant prices. Others use particularly narrow definitions of the disease to ensure that the potential market is extremely small. A few companies have achieved orphan drug status through a 'salami technique': submitting multiple applications for the same drug by specifying different groups of symptoms, each of which affects fewer than 200 000 people.

The possible remedies are several. Applicants could be required to supply medication free to those who cannot afford it; price controls could be introduced; and licensing or co-marketing arrangements could be made mandatory. The drawback of all these approaches is that they undermine the original law's basic objective, which is to provide companies with a special incentive to produce orphan drugs.

---

is discussed in Chapter 4. Chapter 5 takes a detailed look at the industry, examining changes in the cost structure and size of firms. The fact that governments intervene extensively has been noted in this introduction and is fully addressed in Chapter 6. In Chapter 7 attention turns to the types of industry strategies which are currently in use or are beginning to emerge among different sets of producers. The concluding chapter summarizes some of the book's major findings and looks at several likely developments during the 1990s.

NOTES

1. New Zealand is the only exception. That country has invented one NME and has limited research capabilities.
2. The national regulatory bodies that carry out this function usually (but not always) coordinate their work to ensure that names are identical from one country to another. In addition, the World Health Organization (WHO) supplies a set of international non-proprietary names (INN). The confusion which results from the multiple brand names for a single chemical entity is the major professional argument behind WHO's recommendation that physicians prescribe according to the INN.

2. Production and Consumption

Pharmaceuticals is a comparatively small industry in terms of its contribution to national output, employment, exports or income generation. It nevertheless attracts a great deal of attention from consumers and policy makers alike. The industry's true importance stems from the fact that a society's health depends on the availability of modern, efficient drugs at affordable prices. Most of the attention focused on the pharmaceutical industry therefore relates to consumption rather than production.

The first section of this chapter provides a survey of the global pattern of production. The bulk of production occurs in a relatively few countries, but the need for drugs is a universal one. These aspects are considered in the second section where attention is focused on global markets for pharmaceuticals. Given the diverse range of drug requirements, virtually no country is able to meet all its domestic needs. The concluding section of the chapter looks at this aspect, examining the relationship between production and consumption.

GLOBAL TRENDS IN PRODUCTION

Most of today's pharmaceutical firms have existed in one form or another since the nineteenth century. Many of the European companies are offsprings of the chemical industry, where some of the early drug discoveries were made. Several American companies were also satellites of the European chemical industry before they became independent. Others started out as wholesale chemists and makers of patent medicines.

These links suggest a rather long history but in fact the modern pharmaceutical industry did not really begin to take shape until after the Second World War, when pharmaceutical companies turned to modern-day methods of drug development. There were two main reasons for the transformation. First, the rapid growth of demand for newly discovered antibiotics convinced many firms that drug research offered rich rewards. Second, firms turned to the production of speciality drugs once they realised that the market value of products could be protected by patent and promoted through the use of brand names.

Major Producers and Patterns of Growth

Supported by ever-increasing demand for health care, world production of pharmaceuticals has grown at an exceptional pace throughout most of the post-war period. Today nearly 60 countries have annual production levels of at least $100 million. The bulk of world production is nevertheless confined to only a few countries. The leaders continue to be those countries where modern pharmaceutical production first emerged: Belgium, France, the Federal Republic of Germany, Italy, Japan, the Netherlands, Sweden, Switzerland, the United Kingdom and the United States. Only China has managed to develop a substantial pharmaceutical industry during the last two decades.

The global pattern of pharmaceutical production is indicated in Table 2.1. World gross output (in constant 1980 dollars) increased more than twofold between 1975 and 1990. Among the industrialized countries, producers in North America and Japan have expanded most rapidly. Production has stagnated in Eastern Europe and the USSR although in certain countries the industry remains competitive and its prospects are relatively bright (see Box 2.1). Growth in developing countries has generally been slow although performance within this group varies widely. The industry has flourished in certain Asian markets, though in many developing countries production has lagged.

Table 2.2 looks at the pharmaceutical industry in relation to the rest of the chemicals complex and the manufacturing sector. Although growth slowed in the 1980s, output continues to expand at a pace which exceeds that for other parts of the chemicals complex and in comparison with total manufacturers. The pattern is different in developing countries. Production of industrial chemicals and rates of growth of the manufacturing sector as a whole have generally surpassed those of pharmaceuticals, meaning that the industry's relative importance has declined. On average, drug producers account for almost a quarter of all chemical output in industrialized countries but in developing countries the share has declined to 17 per cent (see Statistical Appendix, Table A.2 for details).

Table 2.1   World distribution and growth of production of pharmaceutical preparations, 1975–90ᵃ

Country groupᵇ	Percentage share in world total production		Growth rate (percentage) 1975–90
	1975	1990	
Eastern European countries and USSR	10.2	8.6	4.0
Developed market economies	67.2	73.0	5.8
North America	20.4	22.7	5.9
EC	28.6	24.3	4.1
Other Europe	2.7	2.6	5.0
Japan	14.2	22.3	8.4
Others	1.3	1.1	4.1
Developing countries	22.6	18.4	3.8
Latin America and Caribbean	10.0	7.9	3.5
North Africa	0.5	0.4	3.6
Other Africa	0.8	0.4	0.9
South and East Asia	3.6	4.9	7.3
China	5.6	3.5	2.1
Others	2.1	1.3	1.8
World	100.0	100.0	5.2
World total production (constant US$ billions)	70.1	150.3	

Notes
ᵃ Figures are derived from data on gross output at constant 1980 prices.
ᵇ For the country composition of regions, see Statistical Appendix, Table A.1.

Source: UNIDO, based on data reported by national pharmaceutical manufacturers'
associations; IMS; United Nations trade tapes; national statistical offices.

Box 2.1   The pharmaceutical industry in Hungary

Domestic production of pharmaceuticals in Hungary began in 1901. The industry flourished in the period between the two world wars, with the largest firms having affiliates and joint ventures in Austria, Belgium, Czechoslovakia, Egypt, Greece, Italy, Mexico, Poland, Spain and Turkey. Progress was much slower in the 1950s and 1960s following the industry's nationalization and reorganization. Since 1968, however, the industry has gradually rebuilt its international contacts. Hungarian firms have obtained a number of process patents for the manufacture of new drugs in large volumes. They have also developed a network of international agreements with multinationals. Some of these are licences for the domestic production of patent-protected NCEs. Others are to promote the foreign sale of new and original Hungarian drugs such as Cavinton, Jumex and Osteochin. The innovative capacity and international contacts which the industry has developed have allowed it to remain competitive in several segments of the world drug market and the leading firms were among the first candidates for privatization in 1990. The following table provides some of the main indicators for the industry.

Box table 2.1   A profile of the Hungarian pharmaceutical industry, 1989

Main indicators (US$ millions)		*Structure of income (%)*	
Gross output	860	Pharmaceuticals	77
Value added	333	Pesticides	9
Net income	838	Cosmetics and other[a]	14
Exports	497		
R&D expenditures	56		
Domestic consumption[b]		*Number of employees*	
Local manufacture	218	R&D	4 270
Imports	80	Other	18 830

Notes
[a] Includes licence fees.
[b] At manufacturers' prices.

Source: Union of the Hungarian Pharmaceutical Industry.

Drug Production in Developing Countries

The pharmaceutical industry in developing countries accounts for around a fifth of the world production. The size of firms and pattern of product specialization differ significantly from those in industrialized countries. Most firms are small in size and many of the drugs they produce are branded (usually patent-expired) products. A number of firms in industrialized countries specialize in generic products and most produce their own medicinal chemicals. In contrast, only a few developing countries are engaged in the production of medicinal chemicals. They include Argentina, Brazil, China, Egypt, India, Mexico, the Republic of Korea, Puerto Rico, Turkey and Yugoslavia.

The degree of foreign ownership is another distinguishing feature. In their effort to establish a domestic pharmaceutical industry, a great number of developing countries have turned to multinationals. Policy makers actively encourage foreign investors by offering attractive incentives in the form of tax reductions, generous import allowances, favourable treatment on repatriation of profits and other means. As a result, foreign-owned companies account for about two-thirds of all pharmaceuticals produced in the developing world.

China is the largest producer among the developing countries, accounting for two-fifths of this group's total in 1988. The Chinese industry has evolved along lines that are quite different from those of other large suppliers. Starting from a strong basic chemical industry, the country created a pharmaceutical–chemical industry but did not immediately attempt to move into downstream activities. A processing and dosage-form industry was not added until much later when the country had become relatively self-sufficient in the production of essential pharmaceutical chemicals.[1] Chinese policy makers relied on joint ventures with foreign multinationals when they decided to embark on this phase. The country now operates a number of large, vertically integrated firms that produce both medicinal chemicals and preparations. It also has ample natural raw materials for producing traditional medicines which account for around 20 per cent of total domestic pharmaceutical production.

Today the Chinese industry has sufficient capacity to meet any increases in domestic demand that might result from improvements in the country's national health care systems. The same is not true for many other developing countries. The relatively slow progress of the pharmaceutical industry is partly due to the weak relationships between producers and government authorities and the fact that regulatory controls and policies to promote the industry's development are not well coordinated (see Chapter 6).

Table 2.2 Relative growth and importance of the pharmaceutical industry in terms of value added, by country group, 1975 and 1988[a]

Country group[b]	Relative growth index for the period 1975–88[c]		Percentage share of the pharmaceutical industry in total chemical production[e]		Index of relative specialization within total manufacturing[f]	
	Relative to the growth of total chemical production[d]	Relative to the growth of MVA	1975	1988	1975	1988
Developed market economies	1.60	1.50	18.8	23.9	1.00	1.05
North America	1.58	1.33	17.5	22.1	1.12	1.11
EC	1.71	2.40	18.4	23.6	0.89	1.06
Other Europe	2.48	2.86	16.6	25.1	0.69	0.92
Japan	1.32	0.89	24.5	29.5	1.29	0.99
Others	1.31	2.19	16.5	18.8	0.59	0.68
Developing countries	0.57	0.82	24.7	17.3	1.02	0.78
Latin America and Caribbean	0.37	0.96	26.3	16.6	1.25	1.04
North Africa	1.18	1.17	25.9	29.7	0.79	0.77
Other Africa	1.93	0.93	16.0	19.4	0.70	0.58
South and East Asia	0.71	0.76	23.5	17.1	0.85	0.57
Others	0.96	1.66	19.5	18.7	0.56	0.70
World	1.34	1.37	19.5	22.8	1.00	1.00

Notes

a All figures are based on data in constant 1980 dollars.

b Countries included were those for which relevant data were available: all Eastern European countries, USSR and developing centrally planned economies were excluded. For the comparisons with total chemical production, 69 countries were included, while for the comparisons with MVA, 141 countries were included.

c Relative growth index was defined as the ratio of the growth rate of the pharmaceutical production to that of total chemical production or MVA.

d Total chemical production refers to production of basic chemicals (ISIC 351) and other chemicals (ISIC 352).

e Cross-country weighted averages.

f Index of relative specialization was defined as the ratio of the share of a given country group in the world total pharmaceutical production (value added) to the share of that country group in world total MVA.

Source: UNIDO industrial statistics data base; the United Nations Statistical Office; estimates by UNIDO.

27

These difficulties are most apparent in Latin America. As in other developing countries the pharmaceutical industry is subject to price controls which are far stricter than for other industries, although Latin American rates of inflation have been exceptionally high. Increases in drug prices are allowed only after long intervals and have sometimes been less than the rate of increase in production costs. These effects are compounded by massive devaluations of the national currency. A majority of the region's drug producers import the bulk of their raw materials (as much as 80 to 90 per cent) and these imports become much more expensive as the home currency depreciates.

Changes in government regulations have also limited the demand for drugs. In order to lessen public expenditures, governments in developing countries have reduced their drug lists for public reimbursement and domestic sales of the excluded drugs have fallen. The net effect has been a drop in the profitability of the pharmaceutical industry – both in absolute terms and relative to other industries – and has led to the closure of many drug producers.

Quality assurance is another serious problem for local producers in developing countries. Such assurance is indispensable if the industry is to expand, yet many firms are neither technologically nor financially capable of conducting efficient quality control and testing. The usual reasons are low profit margins and a shortage of highly qualified technical personnel. Such shortcomings can shift consumer preferences towards imports and limit growth of the domestic industry.

The question of quality can be of such importance to consumers that it undermines the government's efforts to improve methods of distribution or reduce prices. For example, when policy makers have taken steps to encourage the entry of local firms supplying generics, the result has sometimes been that the market is flooded by substandard drugs. These products have undergone little, if any, quality control and do not meet minimum standards. Sales of more expensive, branded drugs may actually increase as doctors and patients specify products for which quality can be guaranteed (see Pradhan, 1983, p. 233).

GLOBAL MARKETS FOR PHARMACEUTICALS

Most consumer products have a fairly standardized set of characteristics and attributes wherever they are sold. Although pharmaceuticals fall into this class of goods, drug markets retain several distinctive features. Each medicine is purchased by a limited group of buyers – patients with a particular disease or health problem. Methods of drug distribution are also different from those of other consumer goods. Some drugs are purchased privately while others are distributed through public health care systems and the proportion of total sales handled by each distribution channel varies widely from country to country. Nor are pharmaceuticals sold only to the final user. Over-the-counter (OTC) drugs are marketed as consumer products but prescription drugs are obtained through pharmacies or doctors who act for consumers in determining the types, brands and quantities to be prescribed.

Trends in Consumption

World consumption of pharmaceutical preparations (in constant 1980 US dollars) was $70 billion in 1975 but had more than doubled, to $150 billion, by 1990.[2] During this period the world's per capita consumption of drugs increased by almost 70 per cent, from $17 to $29.

Very few other industries can boast such an impressive growth record, but despite this there is some concern about the industry's prospects. Demand in some therapeutic areas (for example, anti-ulcer drugs) is growing, but in others it stagnates. The slowdown might not be regarded as serious if many new therapeutic drugs were being introduced to the market each year. However a large number of each year's launches are not new products but merely substitutes for older ones. Drug companies have therefore sought to rely on price increases in order to raise total value of sales. The practice helps to raise total sales but it puts the industry in conflict with public officials who wish to reduce the costs of public health care.

Table 2.3 indicates the global pattern of consumption for pharmaceutical preparations in 1975 and 1990. More than 70 per cent of all pharmaceuticals are sold in developed market economies. The developing countries account for less than a fifth,[3] with the remainder being consumed in East European countries and the USSR. Drug usage is growing most rapidly in Japan and North America. The growth of the Japanese drug market has been spectacular. The country's per capita consumption was already the highest of all industrialized countries in 1975 and has continued to rise since then (see Box 2.2). At the other end of the scale are parts of Africa where drug consumption has

Table 2.3 *World consumption of pharmaceutical preparations: regional shares and average per capita consumption, 1975 and 1990 (percentages and 1980 dollars)*

Regional groups[a]	Share in world population 1990	Share in world pharmaceutical consumption		Average per capita consumption of pharmaceuticals (1980 dollars)[b]	
		1975	1990	1975	1990
Eastern European countries and USSR	7.2	10.6	9.3	21.8	37.1
Developed market economies	15.9	65.4	71.7	60.6	130.7
North America	5.4	20.5	23.0	58.6	123.9
EC	6.3	26.0	22.5	57.0	102.9
Other Europe	0.6	2.3	1.8	51.5	85.7
Japan	2.4	15.0	23.0	92.0	276.6
Others	1.2	1.6	1.4	24.4	35.6

	76.9	23.9	18.9	5.7	7.1
Developing countries					
Latin America and Caribbean	8.5	7.7	6.0	16.8	20.3
North Africa	2.8	1.0	0.9	7.4	9.0
Other Africa	9.0	2.0	1.0	4.7	3.3
South and East Asia	32.0	4.8	5.6	2.8	5.0
China	21.7	5.7	3.6	4.3	4.8
Others	2.9	2.7	1.8	18.2	18.0
World	100.0	100.0	100.0	17.2	28.9

Notes
[a] For the country composition of regions, see Statistical Appendix, Table A.1.
[b] Calculated as total consumption divided by total population.

Source: UNIDO, based on the data reported by national pharmaceutical manufacturers' associations; IMS (various issues); United Nations trade tapes; national statistical offices.

Box 2.2   Drug consumption in Japan

The average Japanese consumes twice as many drugs as an American or West European. This extraordinarily high level of consumption is a result of a unique set of policies relating to methods of reimbursement, prescription and other practices, all of which are discussed in Chapter 6. Much of the spending on drugs is financed from public funds. As these costs began to mount, government officials have imposed tighter controls. The Ministry of Health and Welfare (MHW) has lowered reimbursement levels by more than 30 per cent since 1985. It has also cut the prices of 'me too' drugs and other types of copies although producers of innovative drugs are allowed to sell at higher prices – at least for the first few years. Despite these efforts, the Japanese drug market is still the world's most highly priced. Spending levels may grow even higher in the future as the country's population begins to age.

Box table 2.2   Pharmaceutical sales in Japan, 1975–90 (current manufacturers' prices)

Year	Sales	
	Total (US$ bn)	Per capita (US$)
1975	5.72	51
1980	14.40	123
1985	18.93	157
1990 (projected)	31.91	258

Source: *Data Book 1989*, Japan Pharmaceutical Association (1989) and UNIDO projection.

---

decreased and South and East Asia where per capita consumption has been growing but is still very low – around $5 per capita in 1990.

Spending Patterns

Income will have an obvious effect on consumption but there are also other important determinants. These include price trends, characteristics of the distribution system, the age structure of the population and the national system of health care. Each of these factors is examined below.

Spending in developed market economies has clearly risen over time. This fact is confirmed by Table 2.4, which shows that pharma-

Table 2.4   *Pharmaceutical consumption as percentage of GDP, by region, 1975 and 1990*

Regional groups[a]	Pharmaceutical consumption as percentage of GDP[b]	
	1975	1990
Developed market economies	0.65	0.95
North America	0.52	0.87
EC	0.68	0.71
Other Europe	0.43	0.42
Japan	1.15	1.62
Others	0.48	0.46
Developing countries	0.79	0.67
Latin American and Caribbean	0.94	0.72
North Africa	0.77	0.67
Other Africa	0.63	0.65
South and East Asia	0.68	0.60
Others	0.76	0.81

Notes
[a] For the country composition of regions see Statistical Appendix, Table A.1.
[b] Calculated as total consumption divided by total GDP in current dollars.

Source: UNIDO, based on the data reported by national pharmaceutical manufacturers' associations; IMS; World Bank.

ceuticals accounted, on average, for 0.65 per cent of per capita GDP in 1975; by 1990 the figure had risen to 0.95 per cent. Japan reports the largest share of per capita income spent on pharmaceuticals (1.62 per cent in 1990) – almost twice the level in North America and some parts of Western Europe. No similar increases have occurred in developing countries. In fact the share of per capita gross domestic product (GDP) spent on pharmaceuticals has actually declined in many of the poorer countries.

Consumers in rich countries spend heavily on pharmaceuticals and the share tends to rise as income grows. The relationship between income and consumption of drugs is not a straightforward one, however: increases in income will have different implications depending on the product category. Generics, for example, may sometimes be complementary to patented drugs but in other instances they are substitutes. Similar ambiguities arise when products are characterized in terms of distribution channels. Prescription drugs are more expensive than OTCs, but the two are not always substitutes. As consumers grow richer, spending on both product types could rise.

The relationship between income and consumption is further obscured by large international differences in the market shares of certain product categories. Figures for 1990 show that generics accounted for 50 per cent of all drug sales in Denmark, 17 per cent in Germany and 30 per cent in the United States. They have yet to gain a foothold in most other industrialized countries and their share of the national market is negligible. The importance of generics also depends on the size of the public market and the list of reimbursable drugs. This is because policy makers try to encourage the purchase of generics. In the United Kingdom, for example, they account for 35 per cent of National Health Service (NHS) prescriptions but represent only 10 per cent of the total value of drugs sold.

Generics are generally of less importance in developing countries. One reason is that multinationals are the main suppliers and prefer to sell patent-protected drugs whenever possible. Another is that brand loyalty is especially strong among consumers in developing countries: some buyers fear that local imitations are of poor quality; others buy a particular brand because they know the product, its packaging and its effects. The effectiveness of the advertising and promotional campaigns of multinationals and the modest degree of countervailing power which exists in these countries helps to explain these attitudes.[4]

The situation is even more complicated in the case of OTCs. Most of these are generics which sell at lower prices than prescription drugs or patent-protected medicines. However, in developing countries, the shortage of doctors and health professionals means that many prescription drugs are actually sold over the counter. The result is that OTCs probably have a larger share of the market in developing countries than in industrialized ones (see Box 2.3).

Box 2.3   The share of OTCs in national markets

OTCs account for 20–40 per cent of the pharmaceutical market in Switzerland, the United Kingdom and the United States, but are of little significance in other industrialized countries. Less information is available for developing countries, although the figures for Mexico and Saudi Arabia (see below) are probably representative.

World-wide, OTCs are less than 20 per cent of the total pharmaceutical market, though experts predict that by 2005 their share will match that of prescription drugs. Conditions in the EC are most promising. One reason is that markets for OTCs are still highly fragmented, with only one or two specific brands being available in all 12 EC countries. The situation should change with the creation of a single European market. Companies also expect the sale of OTCs to grow since more consumers will resort to self-treatment as reimbursements are scaled back. The development of an OTC market in Europe would still require that consumer buying habits converge and that national codes regarding labelling practices and contents are harmonized.

Box table 2.3   The market share of OTCs in selected countries, latest year

Country	Share in total pharmaceutical sales	Year
Austria	9	1987
Belgium	15	1987
France	19	1987
Germany, Federal Republic of	11	1988
Italy	6	1988
Japan	9	1989
Mexico	25	1985
Netherlands	11	1987
Saudi Arabia	42	1988
Sweden	9	1987
Switzerland	38	1988
United Kingdom	22	1989
United States	23	1988

Source: UNIDO estimates based on IMS and Scrip (various issues).

International Differences in Drug Prices

The prices of drugs do not play an important role in determining consumption patterns in the developed world where public health insurance covers the majority of the population. Figure 2.1 shows relative

The World's Pharmaceutical Industries

Figure 2.1   *Relative prices and per capita consumption of pharma-*
ceutical preparations in current dollars, 1983 (index:
Sweden = 100)

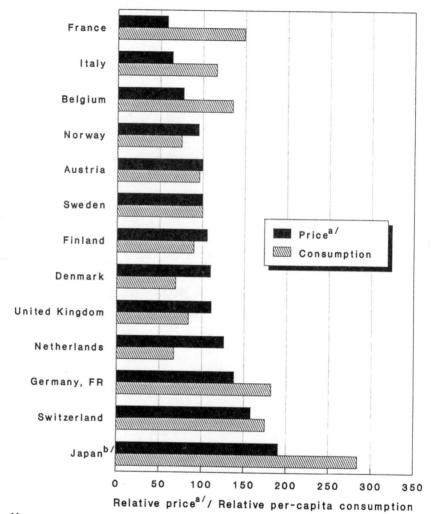

Notes
[a] For selected medicines.
[b] Based on 1982 average price.

Source: Based on data provided in R. Chew, G. Smith and N. Wells, *Pharmaceuticals in Seven Nations, 1985, World Drug Manual, 1988*, and data provided by the Pharmaceutical Manufacturers' Association of Japan.

prices and per capita consumption in several industrialized countries. The range is extremely wide: Japan's relative prices are more than three times greater than those of France. Even within Europe the variation is great. Prices in the Netherlands, Germany and Switzerland are comparatively high, while those in Belgium, Italy and France are low. Per capita consumption is highest at the two price extremes. The Japanese, Swiss, Germans and French consume significantly larger quantities of pharmaceuticals than those in the base country (Sweden). Government regulations are a major reason for these price variations. Several countries have elaborate systems of price control, though others – for example, Canada, Finland and the United States – have few.[5]

Prices are probably an even more important determinant of drug consumption in developing countries since purchasing power is limited and most drug purchases are not reimbursable. A majority of countries impose strict price controls in the public sector and try to ensure that the most essential medicines are available through these channels (see Chapter 6). They allow more freedom for prices in the private sector in order to compensate firms for some of their lost profits. The way in which prices affect consumption is further obscured by the highly segmented nature of drug markets. These factors, together with the limited amount of information which is available, make it very difficult to compare price levels or relative prices in developing countries (see Box 2.4). The following discussion concentrates instead on price trends in public and private markets.

The International Dispensary Association (IDA) publishes price indicators for the case of essential drugs distributed through the public sector. These statistics, in conjunction with other data on sales at the product level, have been used to construct a price index for 46 essential drugs consumed in developing countries. Figure 2.2 shows that the prices declined sharply in the first half of the 1980s. In later years they rose and, by 1990, were 55 per cent greater than in the base year (1985). These results are based on data expressed in United States dollars and may exaggerate the effects of price swings for developing countries which made purchase agreements in currencies other than United States dollars.[6] Another characteristic which is important to note is that the IDA is a non-profit organization which distributes essential drugs at reduced prices. In this sense the index could understate the overall increase in prices. The general impression is that the prices of essential drugs are moderately to substantially

Box 2.4   Variations in drug prices in developing countries

The reasons for the prices of drugs varying so widely between countries are many. Differences in national policy – for example, variations in methods of price control, reimbursement practices and patent systems – are obviously important and are discussed in Chapter 6. The degree of market power enjoyed by the major firms and the internal pricing mechanisms of multinationals and their subsidiaries are other reasons which are also discussed in later chapters. The following table illustrates how great the effects of these and other factors can be. The prices of a particular drug can be 20–30 times more expensive in one developing country than in another.

Box table 2.4   Comparison of prices of drugs in Pakistan and other developing countries (in Pakistani rupees)

Product	Pakistan	India	Sri Lanka	Indonesia
Chloramphenicol 250mg, 12 caps	5.50	9.33	25.66	25.03
Metronidazole 200mg, 10 tabs	3.70	3.75	5.08	59.12
Ferrous sulphate 200 mg, 15 caps	9.00	11.49	17.86	...
Ibuprofen 200mg, 10 tabs	4.83	8.33	14.05	17.39
Propranolol HC1 10mg, 250 tabs	62.70	66.26	152.71	...
Salbutamol 2mg, 10 tabs	3.00	1.51	4.03	13.91
Nifedipine 10mg, 10 caps	6.34	8.15	7.46	27.84
Cimetidine 200mg, 10 tabs	27.59	10.81	23.83	67.81

Source: PPMA, Scrip, World Pharmaceutical News, 1292, p. 18.

higher than in 1985 and therefore represent an added burden on the public health care programmes of many developing countries.

Information on the prices of drugs sold outside the public sector is also scarce but a limited amount of data (which refer mainly, but not exclusively, to drugs distributed through private channels) are shown in Table 2.5. In general, increases in the consumer price index have far exceeded the rise in drug prices. Inflation has been rampant in

Figure 2.2   Price index of essential drugs consumed in developing countries, 1980–90 (1985=100)[a]

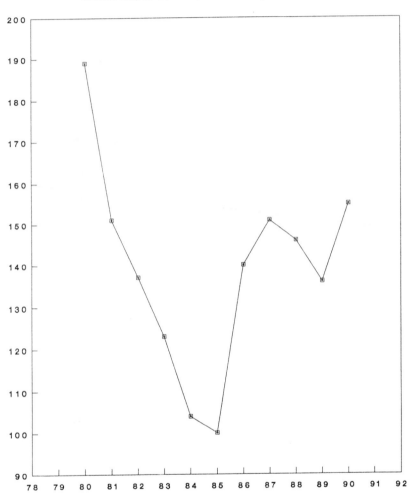

Notes
[a] In estimating this index, therapeutic groups in 20 representative markets were ranked by sales volume. Drugs were then selected on the basis of the ten most important therapeutic groups and the information on sales in each group served as the basis for weighing the prices of individual products. The original prices of the products were converted into dollars using United Nations exchange rates.

Source: UNIDO calculations from IDA Price Indicators, various issues, International Dispensary Association (IDA), Amsterdam, The Netherlands.

The World's Pharmaceutical Industries

Table 2.5   Indexes for consumer prices and pharmaceuticals in
selected developing countries (1980 = 100)

Country	Period	Price indexes Consumer prices	Price indexes Pharma- ceuticals	Country	Period	Price indexes Consumer prices	Price indexes Pharma- ceuticals
Argentina[a]	1984 Sep.	23 600	415	Pakistan	1986	146	136
	1984 Oct.	28 200	499		1987	153	144
	1984 Nov.	32 500	603		1988	166	157
	1984 Dec.	38 900	771		1989	179	171
	1985 Jan.	48 600	979				
	1985 Feb.	58 700	1 198	Peru[c]	1981	175	139
					1982	288	178
					1983	609	578
China	1965	...	151		1984	1 280	1 647
	1978	91	97		1985	3 372	3 624
	1984	109	111				
	1985	122	115				
				Philippines	1981	113	110
					1982	125	120
Ecuador	1981	116	111		1983	137	138
	1982	135	129		1984	206	190
	1983	201	173		1985	254	206
	1984	264	233		1986	256	233
	1985	337	275		1987	265	252
	1986	415	349				
	1987	538	453				
				Yugoslavia[c]	1981	140	135
					1982	184	168
India	1981	113	110		1983	258	223
	1982	122	125		1984	399	327
	1983	136	138		1985	687	668
	1984	148	139		1986	1 304	984
	1985	156	144				
	1986	170	145				
	1987	184	159				
	1988	202	187[b]				

Notes
[a] Drug price index, October 1983 = 100.
[b] June 1988.
[c] Implicit deflator of pharmaceutical production.

Source: Based on data obtained from UNIDO questionnaires to national statistical office,
IMS and Scrip. Exchange rates are from the International Monetary Fund (IMF).

several of these countries, however, and it is unlikely that the money incomes of poorer households have kept pace. For this income class the relative cost of pharmaceuticals has probably grown. Drug companies would also have suffered since their price increases probably lagged behind the rise in production and costs.

These trends are mainly a reflection of domestic policies (primarily price controls). The method of setting appropriate prices and granting price increases can have unintended consequences – particularly if the country is experiencing hyperinflation combined with a steady devaluation of its currency. The problem is illustrated by the Brazilian data in Figure 2.3. Pharmaceutical prices rose by less than the rate of inflation in most years. Meanwhile the position of firms which were large importers of intermediate inputs deteriorated as the value of the cruzeiro fell.

Demographic Effects

As the population of a country ages, the need for drugs grows. Projections show that the population in many industrialized countries is aging quickly. The proportion of those over 45 years old will increase from a third of the total population to more than 40 per cent by the year 2015. In comparison, only 25 per cent of the developing countries' population will be over 45 years of age by 2015.

The relation between age structure and drug consumption is illustrated by the data in Table 2.6. The table shows the four largest selling pharmaceutical preparations in each of 30 countries. In industrialized countries where the proportion of elderly is relatively high, the pattern of drug consumption reflects the high incidence of chronic or degenerative diseases (for example, heart diseases, diseases of the circulatory system and rheumatism).

The situation is much different in developing countries. Because the population is relatively young, the most common diseases are either acute or infectious. Systemic antibiotics, systemic antirheumatics, vitamins, and cough and cold preparations are the most important therapeutic categories in terms of sales. Fortunately the drugs used to treat these ailments are less expensive per patient day than those required for the treatment of chronic diseases.

The frequency of drug consumption also changes with the age structure. Table 2.7 uses data for the Federal Republic of Germany to

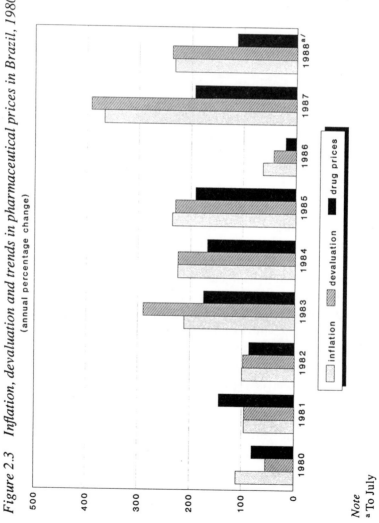

Figure 2.3 Inflation, devaluation and trends in pharmaceutical prices in Brazil, 1980–8
(annual percentage change)

Note
[a] To July

Source: Scrip, *Yearbook,* 1989.

illustrate this characteristic. Roughly 55 per cent of the population aged over 44 years takes drugs at least once a week but only 15 per cent of those between 14 and 44 years of age consume drugs this frequently. Moreover the types of drugs required by individuals aged under 45 years are different from those purchased by the elderly. They consist mainly of pain-killers, cough and cold preparations and digestives. These are frequently generics and are less expensive than other drug types.

Pricing practices can reinforce the effects of population aging. About one-quarter of the average increase in drug expenditures of the statutory health insurance scheme in the Federal Republic of Germany was attributed to the application of new or improved drugs during the period 1975–82. The comparable figure for 1965–75 was only 16 per cent (PMAG, August 1983, p. 57). The difference suggests that the prices of new drugs tend to be much higher – a practice which may only partially reflect the rising costs of R&D.

Systems of Public Health Care

Apart from price and non-price controls, governments influence the pattern of consumption through their national health care systems. The systems in industrialized and developing countries differ greatly and their effects on the pattern of consumption are equally disparate.

Governments in industrialized countries established very generous systems of health care in periods when national income was rising and the proportion of old people in the population was comparatively small. Table 2.8 shows that the public sector accounts for more than half of all drug expenditures in most industrialized countries. The only exceptions are Canada and the United States; in both countries a large portion of pharmaceutical expenditures is covered by private medical insurance or involves direct payments by patients.

Most industrialized countries are now being forced to cut back on their health care programmes and pharmaceuticals are a favoured target. Reasons for the cutback include the rising costs of caring for an aging population, the introduction of new and more expensive drugs and a general tightening of federal budgets. Because a large portion of the total spending on drugs is paid for through public funds, the change in policy is having a dramatic effect on consumption. The steps taken by policy makers include the introduction of

*Table 2.6   Shares of top four drugs in total domestic sales in
selected countries, 1987 (per cent)*

Country	Systemic antibiotics	Antacids anti-ulcerants	Systemic anti-rheumatics	Cerebral peripheral vasodilator	Analgesics
Developed market economies					
Australia	7	4	...	...	5[a]
Austria	11	...	3	5[b]	...
Belgium	8	...	4	...	...
Canada	6	5[e]	5	...	...
France	5	...	...	7	4
Germany, Fed. Rep. of	5	4	...	5	...
Ireland	6	5	5[a]	...	4
Italy	12	6	4	4	...
Netherlands	8	6	...	...	...
New Zealand	...	4[f]	5	...	...
Portugal	11	5[h]	6	4	...
South Africa	7	...	4	...	4
Spain	8	...	...	5	...
United Kingdom	8	7	8	...	...
United States	9	5	...	...	4

Country	Systemic antibiotics	Vitamins	Systemic anti-rheumatics	Cough & Cold preparations
Developing countries				
Argentina	9	...	5	4[h]
Brazil	8	...	4	4[j]
Colombia	13	...	5	...
Costa Rica	9	6	...	4
Ecuador	16	5	6	3[m]
Egypt	16	6	5	...
Indonesia	13	3	...	...
Mexico	11	3[a]	3	...
Pakistan	17	7	...	4
Peru	16	3	4	6
Philippines	18	6	...	9
Republic of Korea	13	6	...	6
Saudi Arabia	8	...	4[f]	...
Turkey	19	4	...	...
Venezuela	11	6	4	...

Notes
[a] Including anti-asthmatics.
[b] Including cardiac therapies.
[c] Blood perfusion solutions.
[d] Including analgesics.
[e] Including sex hormones.
[f] Including analgesics and sex hormones.
[g] Including systemic antibiotics.

Anti-asthmatics	Beta blockers	Psycholeptics	Cough & cold preparations	ACE inhibitors	Cardiac therapies	Others
...	...	...	...	4	...	...
...	...	...	...	...	...	7[c]
...	...	6	...	...	4[d]	...
...	4	...	...	...	...	...
...	...	...	...	3	...	...
...	...	...	...	...	6	...
...	...	...	...	...	...	...
...	...	...	...	...	...	...
7	6	...	...	...	...	...
11	6[g]	...	...	...	...	...
...	...	...	...	...	...	...
...	...	...	4	...	...	...
...	...	...	6	...	...	7[i]
7	...	...	...	...	...	...
...	...	4	...	...	...	...

Analgesics	Antacids anti-ulcerants	Other
...	...	5[d]
...	3	...
4[k]	4	...
3	...	...
...	...	...
4	...	...
2	2	...
4	...	...
5	...	...
...	...	...
5	...	...
...	...	9[n]
3	...	2[o]
4	4	...
5	...	...

[h] Including psycholeptics.
[i] Dietetics.
[j] Including analgesics and vitamins
[k] Including dietetics and vitamins.
[m] Including gynaecological anti-infectives.
[n] Tonics.
[o] Cerebral peripheral vasodilator.
Source: IMS, World Drug Market Manual 1988.

Table 2.7   Frequency of drug consumption by age group in the Federal Republic of Germany, 1990 (percentage)

No. of persons	Total 1027	14–34 years 355	35–44 years 177	45–64 years 341	65 years and over 154
Daily or almost daily	23	9	11	28	55
Once or several times a week	9	6	3	12	15
Once or several times a month	28	32	36	23	18
Seldom or never	41	53	49	37	12

Source: *Pharma Daten 90*, Pharmaceutical Manufacturers' Association of the Federal Republic of Germany, Frankfurt am Main, September 1990, p. 40.

stricter price controls, the withdrawal of reimbursements from selected products and increases in patient contributions to the cost of drugs.

Japan, with the highest per capita spending on pharmaceuticals in the world, has long been an exceptional case. In the last decade, however, the government took aggressive steps to slow the growth of drug consumption. Between 1980 and 1987 the average reimbursement price for drugs in Japc was reduced by 40 per cent (IMS, 1988). As a result, drug oonsumption (measured at current domestic prices in yen) rose by only 4.7 per cent during 1983–6 compared with a 48.6 per cent increase in the previous three-year period.

Doctors in several industrialized countries are now coming under pressure to prescribe more economically and to use generic names so that pharmacists can dispense the cheapest available preparations. Some countries have consciously begun to encourage 'generic substitution' in order to reduce pharmaceutical costs. That tactic is most popular in the United States. Generics accounted for 29 per cent of all the country's drug sales in 1988 and their share is expected to rise to more than one-third by the early 1990s. One reason for the rapid

Table 2.8   Share of public drug expenditure in total national drug
expenditure in selected countries in the mid-1980s (per
cent)

Country	Year	Share of public drug expenditure	Country	Year	Share of public drug expenditure
Australia	1985	54	Italy	1986	64
Austria	1985	75	Netherlands	1986	62
Canada	1985	23	Norway	1986	60
Denmark[a]	1988	63	Spain	1986	75
Finland	1988	68	Sweden	1986	78
France	1986	63	Switzerland	1987	100
Germany, Fed. Rep. of	1985	63	United Kingdom[b]	1987	78
Ireland	1985	70	United States	1985	10

Notes
[a] Including prescription fees.
[b] Public expenditure refers to the National Health Service (NHS) only.

Source: National pharmaceutical producers' associations; IMS (various issues).

acceptance of generics is that a majority of the American population
depend on private medical insurance. Another is that legislation fa-
vours this method of distribution. For example, laws in some states
allow the pharmacist to fill prescriptions with generics unless the
doctor has specifically indicated that a branded product be dispensed.[7]

Generics are not so widely used in Western Europe. The slow rate
of acceptance is partly due to opposition from industry representatives and from the medical profession. Both groups have vigorously
resisted legislation to promote generic substitution.

Systems of public health care are a less significant determinant of
consumption in developing countries. The main reason is the small
portion of total income available for this purpose. Public expenditures
on health care in most developing countries account for between one
and two per cent of gross national product (GNP), while in industrial-

ized countries the figure is much higher, typically between six and eight per cent. As much as two-thirds of the drugs purchased by patients in poor and medium-income developing countries are paid for privately (Redwood, 1987, pp. 254–6.)

Studies carried out in the mid-1980s suggest that the demand for public health care becomes apparent only after a country reaches a certain minimum level of income. For example, in 12 countries with a per capita GNP of less than 500 dollars, public health care expenditures averaged one dollar per capita. The average was $5 in another 18 countries with per capita incomes ranging between $500 and $1000 and reached $54 dollars in a set of 13 countries where per capita income exceeded $1000. In contrast, public health expenditures averaged $376 per capita in a sample of industrialized countries.[8]

Evidence that present systems are fragmented and incomplete is found in a survey of 37 Commonwealth countries. The results showed that the average number of pharmacies per million inhabitants was 27, while the average number of pharmacists was 67. Comparable figures for industrialized countries are much higher. For example, the average number of pharmacies in Australia, Canada, New Zealand and the United Kingdom was 252 per million inhabitants, while the average number of pharmacists was 651 (Commonwealth Secretariat, 1985, p. 14–15).

The per capita amounts spent on drugs are small but the general shortage of medical facilities in developing countries means that they account for as much as a third of all public health care expenditures.[9] In industrialized countries the share is much smaller, around 5–15 per cent, or between $20 and $60 per capita.

Like their counterparts in industrialized countries, policy makers in some developing countries are promoting the use of generics to cut costs (see Box 2.5). Such efforts are not part of a general trend, however. The market share of generics is still small and government officials face substantial resistance in attempting to encourage the replacement of patented and brand-name drugs by generics (see Chapter 6).

The interrelationship between all these determinants is complex, but at the global level the net effect seems to be a slowdown in growth of consumption. Nominal growth in consumption of pharmaceutical preparations (expressed in national currencies) fell in 15 of 23 industrialized countries in 1983–6, compared with 1980–3. Among the developing countries, 49 out of 97 countries experienced a similar

---

Box 2.5   Promoting generic drugs in developing countries

In 1988 the government of the Philippines introduced a policy to promote and encourage the use of generic names so that patients could buy cheaper equivalents. The policy applies to all phases of firms' operation – manufacturing, marketing, advertising – and also to the prescription of drugs by the doctors. Pakistan and Bangladesh have passed laws similar to that in the Philippines. Argentina, too, has indicated that the introduction of compulsory generic prescriptions is a long-term health care objective.

A more comprehensive approach has been taken in Nigeria. In 1991 all pharmaceutical products available in the country were deregistered. Manufacturers and importers were then permitted to reregister those included in the essential drug list (EDL). Products not included in the EDL cannot be imported, manufactured or sold in Nigeria. The purpose of the new law was to encourage the local production of essential drugs. In practical terms, generics now account for the entire drug market in Nigeria.

---

decline. The deceleration may be only a short-term aberration resulting from trends in prices, incomes and government spending. Whatever the causes, changes in consumer behaviour carry important implications for the relationship between production and trade. This aspect is examined below.

THE LINKS BETWEEN PRODUCTION AND CONSUMPTION

The pharmaceutical industry in most countries is geared mainly to meet domestic demand. The need for drugs is a universal one, however. This fact, along with the highly concentrated pattern of world production, means that trade is substantial. The following discussion begins with an examination of the relationship between exports and production and then considers the links between imports and consumption. Finally it assesses the role of domestic and foreign demand in the growth of exports.

Share of Exports in Production

The relative importance of exports depends on the size of the country, the type of drugs produced and the home industry's technological capabilities. Not surprisingly, there are wide differences among the world's producers. According to Table 2.9 the export/production ratio is highest among pharmaceutical companies in Western Europe and has been rising since the mid-1970s. In contrast, North America exports comparatively little of its production. The average is much higher, however, when only the larger producers of pharmaceuticals

Table 2.9   Share of exports in production of pharmaceutical preparations,[a] 1975 and 1989

Country group[b]	Cross-country weighted averages of the percentage ratio of exports to production[c]	
	1975	1989
Developed market economies	*12.1*	*10.1*
North America	3.1	2.3
EC	20.6	24.0
Other Europe	57.8	68.1
Japan	0.6	0.3
Others	7.3	8.3
Developing countries[d]	*2.6*	*6.7*
Latin America and Caribbean[d]	2.2	2.8
North Africa	1.8	6.7
Other Africa	1.2	3.0
South and East Asia	4.8	5.5
China	1.1	4.2
Others	5.2	32.3

Notes
[a] Based on data in current dollars.
[b] For the country composition of regions, see Statistical Appendix, Table A.1.
[c] Intra-region group trade is included. For several countries exports might include a significant amount of re-exports.
[d] Excluding Puerto Rico.

Source: see Table 2.1.

are considered. Big American companies receive about 30 per cent of their revenues from overseas markets, while in Western Europe the figure ranges between 60 and 95 per cent.

The Japanese industry produces almost entirely for domestic consumption. On average it exports less than one per cent of total production and even the largest firms have an export/production ratio of only around 6 per cent. This pattern now seems to be changing. Japanese producers traditionally relied on foreign firms to get their products approved and distributed abroad. They have now begun to try to guide their own drugs through the American and European systems of approval. Often they do this through joint ventures and licensing arrangements with firms based in these markets (see Chapter 7).

On average exports account for less than 6 per cent of production in developing countries. First, policy makers usually insist that local producers give priority to the domestic consumers and cater to the domestic disease pattern. Second, the main purpose of multinational investment in developing countries has been to maintain or expand their market share rather than to reduce production costs or establish an export base. Third, international marketing of drugs is more difficult than for other consumer goods, mainly because of the many forms of government intervention and the high standards set for products. Finally, the production of pharmaceutical preparations (excluding some traditional drugs) does not depend on cheap labour or resource abundance and few of these firms would have a cost advantage in international markets.

Foreign Supplies of Pharmaceuticals

Though exports are only of modest importance in most countries, imports often account for a significant portion of drug supplies. The extent to which a country meets its home demand is roughly indicated by the ratio of domestic production to consumption.[10] Obviously certain drugs will be produced in excess of home demand, while domestic requirements for others exceed production. In other words, a country that is self-sufficient may still have a moderate-to-large amount of imports. This latter characteristic can be represented by the import/consumption ratio.

Table 2.10 gives regional averages for both ratios. North America, China and Japan, are self-sufficient and depend very little on imports. West European countries are also self-sufficient but imports are still

Table 2.10 Self-sufficiency and import dependency of the industry of pharmaceutical preparations, 1975 and 1989, by country group (per cent)

Country group[a]	Ratio of production to consumption		Ratio of imports to consumption[b]	
	1975	1989	1975	1989
Developed market economies	104.8	102.1	7.9	8.2
North America	102.3	99.5	0.9	2.7
EC	111.9	108.7	11.1	17.5
Other Europe	115.8	141.4	51.1	54.8
Japan	97.0	98.1	3.6	2.1
Others	80.9	75.7	25.0	30.5
Developing countries[c]	86.4	85.9	15.8	19.8
Latin America and Caribbean[c]	93.7	92.0	8.3	10.6
North Africa	52.4	44.5	48.5	58.5
Other Africa	41.6	40.0	58.9	61.2
South and East Asia	80.1	89.8	23.8	15.1
China	100.4	100.5	0.7	3.7
Others	80.4	76.6	23.8	48.2

Notes
[a] For the countries which were included in the calculations, see Statistical Appendix Table A.1.
[b] Intra-regional trade is included.
[c] Excluding Puerto Rico.

Source: See Table 2.1.

large relative to consumption. No developing regions other than China have levels of production that are roughly equivalent to their drug needs. Many countries in Africa depend mainly on imports. In South and East Asia the degree of self-sufficiency has risen slightly, while in Latin America it has declined.

The averages in Table 2.10 are based on data for 139 countries, but only 17 are net exporters of pharmaceutical preparations (see Statistical Appendix, Table A.1). Most of the surplus countries have small home markets and a high degree of intra-industry trade. Switzerland is an extreme case. It produces 2.7 times more than its domestic requirements and 80 per cent of its production is exported. Nevertheless half the pharmaceuticals consumed in that country are imports. In contrast countries with large home markets have a different sort of production/trade relationship; they are virtually self-sufficient and have only a moderate or negligible need for imports.[11]

Sources of Growth

The foregoing relationships can be examined more systematically in terms of a 'growth accounting' framework. According to this method, growth is attributed to one of the three broad sources: domestic demand, export expansion or import substitution. The growth contribution of each of the three components is then measured with the help of accounting identities.[12] The approach compares actual changes in each component with hypothetical values which would have occurred if exports, imports and domestic production had grown at the same rate as domestic consumption.

Table 2.11 shows the results of these calculations for 63 countries. Three distinct types of growth paths can be identified. In one group are the larger industrialized countries, where production gains are primarily attributable to the growth of domestic demand. A second group is made up of smaller industrialized countries with a relatively large pharmaceutical industry. The industry's expansion in these countries was due to a combination of domestic demand and exports.

The third group consists of developing countries and small industrialized countries without a significant pharmaceutical industry. Growth in these countries depended on domestic demand and/or import substitution. The domestic market is small and export expansion was an important growth contributor in very few instances. The pharmaceutical industry in these countries depends on the production of multinational affiliates and, to a lesser extent, the licensed production of generic products. In either case, domestic supply is met without heavy local investment in product development and large-scale operations are not required.

In conclusion, very few developing countries have been able to launch any export drive; rather, as consumption grows and drug requirements become diverse, the import dependency of developing countries has risen. Unlike the case of other industries, no form of

Table 2.11 Sources of the production growth in selected countries, 1975–89

	Total incremental output[b] (current $ million)	Source of growth[a]		
		Domestic demand expansion	Export expansion	Import substitution
		(percentage of total incremental output)		
Large developed market economies[c]				
Australia	658	108.1	8.4	-16.4
Canada	2 585	102.7	2.9	-5.6
France	7 566	89.4	21.9	-11.3
Germany, Federal Republic of	8 592	87.4	17.5	-5.6
Italy	6 322	99.3	6.3	-4.3
Japan	49 058	98.1	0.2	1.6
South Africa	464	99.0	-0.2	1.2
Spain	2 088	101.5	8.8	-10.3
United Kingdom	4 276	77.2	40.1	-17.3
United States	35 812	100.0	1.9	-2.0
Small developed market economies[d]				
Austria	387	81.8	17.4	0.8
Belgium	1 063	50.2	45.1	4.7
Denmark	686	32.9	66.9	0.2
Finland	287	74.4	14.2	11.4
Greece	290	87.3	15.1	-2.4
Ireland	491	7.1	19.5	73.4
Israel	115	108.3	26.8	-35.1
Netherlands	728	66.1	45.7	-11.8
New Zealand	115	36.3	1.5	62.2
Norway	124	69.0	8.0	22.9
Portugal	426	85.9	3.6	11.4
Sweden	1 137	32.7	43.3	24.0
Switzerland	2 006	29.6	74.3	-3.9
Large developing countries and areas[c]				
Algeria	54	69.4	0.1	30.6
Argentina	514	101.1	1.2	-2.3
Bangladesh	67	113.6	0.0	-13.6
Brazil	1 316	99.4	2.8	-2.2
China	1 452	98.5	10.0	-8.4
Colombia	293	98.2	0.7	1.1
Egypt[e]	149	178.7	5.8	-84.4
India	1 417	87.4	10.7	1.9
Indonesia	266	92.4	1.2	6.5
Iran (Islamic Republic of)	490	25.5	0.0	74.5
Iraq	137	155.4	0.0	-55.4

	Total incremental output[b] (current $ million)	Source of growth[a]		
		Domestic demand expansion	Export expansion	Import substitution
		(percentage of total incremental output)		
Kenya	27	81.0	-1.5	20.5
Malaysia	55	41.3	4.5	54.2
Mexico	1 051	102.7	0.5	-3.2
Morocco	143	78.3	4.2	17.5
Nigeria[f]	195	40.3	0.0	59.7
Pakistan	322	101.3	0.6	-1.9
Peru[e]	42	137.3	-2.3	-35.0
Philippines	315	114.6	1.7	-16.4
Republic of Korea	1 779	100.8	0.8	-1.6
Sudan	96	102.0	0.1	-2.1
Taiwan Province	628	93.4	1.6	5.0
Thailand	245	73.5	1.9	24.6
Turkey	327	135.4	7.2	-42.6
Venezuela[e]	158	110.1	-0.1	-10.0
Yugoslavia[e]	372	74.4	24.3	1.3
Small developing countries and areas[d]				
Bolivia[f]	20	21.0	0.0	79.0
Chile	68	141.4	2.0	-43.4
Costa Rica	30	61.8	32.7	5.5
Dominican Republic	31	26.5	0.4	73.2
Ecuador[f]	29	85.8	-1.8	16.0
El Salvador	33	94.8	15.5	-10.3
Ghana	54	49.5	0.2	50.4
Guatemala	63	59.7	28.6	11.7
Hong Kong	58	95.6	8.1	-3.7
Jordan	54	0.9	1.1	98.0
Singapore	45	17.3	29.8	53.8
Syrian Arab Republic	67	1.7	0.0	98.3
Tunisia	39	23.4	0.0	76.6
Uruguay	74	89.1	2.3	8.6

Notes

[a] For a definition of the equation used to estimate the contribution of each source of growth, see endnote 12.

[b] Output is measured in gross terms and refers only to pharmaceutical preparations.

[c] Countries/areas with population exceeding 15 million in 1987.

[d] Countries/areas with population of less than 15 million in 1987.

[e] 1975–87.

[f] 1975–86.

Source: See Table 2.1.

intra-industry specialization between developing and industrialized countries has emerged. Thus the pattern of world trade in pharmaceuticals is somewhat distinct. Its supply-side determinants are examined in more detail in the following chapter.

NOTES

1. A major reason was that much of the original technology was transferred from the Soviet Union, which also lacked a processing industry.
2. National data on consumption were compiled for 143 countries/areas (see Statistical Appendix, Table A.1). Unless otherwise noted, 'world' refers to these 143 countries/areas.
3. Important markets by sales (current dollars at manufacturers' prices) in developing countries in 1990 are China ($4.2 billion), Brazil ($2.9 billion), the Republic of Korea ($2.5 billion), Mexico ($2.0 billion), India ($1.8 billion), and Argentina ($1.2 billion).
4. Consumers are very poorly organized in developing countries, methods of controlling advertising and other practices are often lax, and many doctors have only a limited awareness of OTCs and non-branded products.
5. The extent to which drug prices are controlled can be seen in terms of the relative growth of these prices. In France, controls are fairly stringent. The price index of reimbursable drugs in that country rose by only 24 per cent between 1980 and 1988, although the consumer price index increased by 72 per cent. The pattern was different in the Federal Republic of Germany, where price controls are more lenient. The German pharmaceutical price index increased by 24 per cent in 1980–8 compared with a 22 per cent rise in the consumer price index (IMS, 1990).
6. The IDA reports its prices in Dutch guilders and when the index is expressed in this currency long-term movements in prices are much less volatile.
7. The laws have a significant impact since the pharmacists' profit is generally higher on branded products than on generics.
8. The industrialized countries referred to here are Australia, Canada, New Zealand and the United Kingdom. See Commonwealth Secretariat, 1985, pp. 14–15 and IMS, 1988.
9. For instance, in Argentina, pharmaceutical expenditure accounted for 32 per cent of total public health care expenditure in 1985 (IMS, 1988, p. 1.14).
10. The measure ignores annual changes in stock. Thus the relationship between domestic production and consumption is equivalent to the difference between exports and imports. It should be noted that the levels of consumption in many developing countries may be much lower than the levels of potential demand because of their limited capability to import.
11. To test these relationships Spearman rank correlations were calculated using 1987 data for over 80 countries (see Statistical Appendix, Table A.3). Both the

degree of self-sufficiency and trade dependence are highly correlated with population (a proxy for domestic market size) and carry the expected signs. The two measures are only weakly associated with income (measured by per capita GDP). Furthermore the ratio of exports to production – a measure of export dependency – is negatively related to population, while its association with GDP per capita is strongly positive. All these results suggest that only large countries can be self-sufficient without depending on trade. Small countries may be self-sufficient but must still depend heavily on trade (both imports and exports).

12.   The relationship can be expressed as follows:

$$(Q_2 - Q_1) = d_1 (C_2 - C_1) + d_1 (X_2 - X_1) + (d_2 - d_1)(Q_2 + M_2),$$

where Q = domestic production, C = consumption, X = exports, M = imports, d = ratio of domestic production (that is, gross output) to total supply $(Q + M)$ and the subscripts 1 and 2 refer to 1975 and 1989 respectively. The first term on the right-hand side of the equation represents the growth in domestic demand while the second term represents export expansion. The third term measures the difference between actual imports in 1989 and the hypothetical level of imports that would have occurred if the import share in total supplies had remained equal to that in 1975.

3. Competitive Trends in International Markets

The pharmaceutical industry is clearly passing through a period of considerable change. The possibility that these changes will alter the competitive position of producers in different countries is especially important, eliciting a variety of responses. Some commentators have suggested that Japanese producers will soon emerge as world leaders, mainly at the expense of drug firms in the United States and Western Europe. Others see the issue of competitive shifts in a different light, arguing that the extent of government intervention in some countries is so great that it stifles innovation and progress. In response government officials point to studies showing that many firms continue to thrive in international markets despite extensive government regulation at home (see, for example, Burstall and Senior, 1985).

Several industry-wide trends are examined in this chapter. The results provide useful insights about shifts in competitive abilities, though they cannot provide a comprehensive picture of global trends. One reason is that any assessment of competitive performance can be no more than an informed guess; the concept is an elusive one which is not readily captured in empirical terms. Another problem is the lack of relevant information. Most drug companies are extremely secretive and the markets they serve are highly fragmented. A third complication results from differences in national policy. These variations are often so great that they distort any comparisons between producers in different countries.

Despite these drawbacks, various types of information can be assembled to gain some impression of international competitive standing.[1] An examination of trading patterns is the usual starting-point for such an exercise. The fact that only 30 firms account for nearly half of world sales underlines the importance of foreign markets. The industry itself possesses several characteristics that should encourage trade. First, the high costs of research and the limited period of patent protection for most drugs make it essential that the innovator sells its products on a global scale. Second, there are significant economies of scale in the distribution of medicines and this, too, requires that producers adopt an international outlook. Finally, transport costs are a natural barrier to trade for most manufactured goods but matter very little in this instance since medicines are neither heavy nor voluminous.

At the same time there are factors which inhibit (rather than encourage) trade. One is the array of controls and regulations employed to ensure that standards of safety and efficacy are met. Price controls and restrictions on the availability of medicines are also common. Finally there is the normal range of trade restrictions which include tariffs, quotas, various types of non-tariff barriers and other policies that make it difficult for firms to export. As a result the actual pattern of trade may be quite different from that which would be expected if competitive forces were the sole determinant.

Many of the industry's larger drug producers have responded to these circumstances by locating parts of their operation in foreign markets. The global network of pharmaceutical facilities is now huge, suggesting that investment – in particular, foreign direct investment (FDI) – is an important indicator of competitive abilities. Unfortunately, data on FDI are limited. For that reason, much of the following discussion makes use of information on international sales and total investment.

The nature of competition within the industry suggests a third aspect with implications for the international pattern of competitiveness. Product development rather than low prices has been the major determinant of market leadership, at least in the industrialized countries. Successful firms must always have a number of promising drugs in their research pipeline if they hope to thrive in today's markets. Thus, while trade and investment help to shape patterns of competition, the pace of innovation is the major determinant of competitive leadership in the pharmaceutical industry.

To summarize, international trade and international production (or FDI) are the two conventional ways for firms to reach foreign markets, while a strong innovative capacity is the prerequisite for competitive leadership in major markets. Each of these features is examined below.

EXPORT PERFORMANCE IN THE PHARMACEUTICAL INDUSTRY

Pharmaceuticals are traded in both their intermediate and finished form. Intermediate products or medicinal chemicals are supplied by producers of fine chemicals to processors where they are turned into pharmaceuticals. The processor then exports the finished product to an importer who sells the drug in his home market.

Trade in these two categories occurs both in open international markets and through closed business channels. Most medicinal chemicals are sold in the open market. These are mainly commodity fine chemicals and patent-expired specialty chemicals, both of which are used to produce generic products.[2] Multinationals, on the other hand, sell medicinal chemicals and pharmaceutical preparations to their affiliates, licensees and appointed distributors. In that case, parallel imports (see Box 3.1) and generic drugs are the only sources of price competition.

The available data do not permit a distinction between sales in an open market and those passing through closed business channels. They do, however, allow for trade to be divided into medicinal chemicals and pharmaceutical preparations and this breakdown is shown in Table 3.1. The most obvious feature is the extent to which trade is dominated by the industrialized countries. Producers in Western Europe supply almost three-quarters of world exports while North America accounts for another 12 per cent. Other exporters, including Japan and the developing countries, are of little importance. The figures also reveal a similarity in the composition of pharmaceutical trade. Roughly two-thirds of all pharmaceutical exports are preparations or finished products. Thus a processing industry capable of converting medicinal chemicals into finished products seems to be essential if a country is to be a major exporter of drugs.[3]

Table 3.2 provides a more detailed look at the pattern of exports for a large number of countries. These figures reconfirm the two characteristics noted above. The 15 largest suppliers accounted for 90 per cent of all pharmaceutical exports in 1988, though only one developing country, China, was among this group. Furthermore the major suppliers are exporters of pharmaceutical preparations rather than medicinal chemicals.

Box 3.1   Parallel importing in the pharmaceutical industry

One consequence of policy disharmony is that suppliers often resort to 'parallel importing', buying in markets where the price is low and reselling in others where prices are high. Parallel imports are common in Europe, owing to widespread control of product prices and a lack of harmonization of pricing policies. Unpublished estimates put the EC's market for parallel imports at $500 million per year. Trade of this type has little or nothing to do with international differences in competitive abilities.

Parallel imports can help to smooth out international price differences, but the drug industry is unconvinced of this virtue. It argues that in the EC parallel importers take unfair advantage of a market that is closely regulated. The importers respond that drug manufacturers are using the regulations to stifle competition.

To sell a drug in the EC a parallel importer must obtain a product licence. For this the importer must first provide a European Community Market Authorization (ECMA) to prove that the drug is made in the EC. Approval of such a licence should require 45 days but in practice the period varies widely. In the Netherlands it requires three months and in the United Kingdom 19 months. The importers believe that this is because regulators must get information from reluctant drug manufacturers but a confidentiality agreement between the regulators and drug producers obscures the licensing process.

Drug producers have several means to delay approval of parallel imports. If they change the colour or shape of pills, or switch from tablets to capsules, a new ECMA number must be issued and imports can be delayed. Drugs can also be sold under different brand names but regulators prohibit the import of drugs with 'foreign' names, even though the ingredients are identical. A name change can therefore deter competition from parallel importers.

The fact that there can be an eight-fold difference in the price of a drug in EC member countries creates great incentives for parallel imports. The industry argues that the situation is due not to profiteering but to the fact that national governments often set prices.

It is clear that a small minority of industrialized countries dominate world markets. Only four – the Federal Republic of Germany, Switzerland, the United Kingdom and the United States – accounted for more than half of world exports in 1988. These countries owe their success to a handful of large, integrated firms which operate as multinationals. The multinationals' ability to market their products on an international scale goes a long way towards explaining the concentration of world suppliers, but it is not the only reason for such a pattern. Another is the nature of competition in the industry which depends mainly on the introduction of new products rather than differences in

Table 3.1   Pharmaceutical exports by region and country group, 1975–88

	Value of exports (millions of current US$)			Share in world 1988 (%)	Export composition, 1988	
	1975	1980	1988		Preparations[a] as a share of total (%)	Medicinal chemicals[b] as a share of total (%)
Developed market economies	5 730	12 056	23 538	88.8	66.1	33.9
North America	801	1 628	3 266	12.3	48.1	51.9
EC	3 650	7 588	14 196	53.5	72.2	27.8
Other Europe	1 095	2 437	5 156	19.4	68.2	31.8
Japan	125	299	734	2.8	22.3	77.7
Others	60	105	186	0.7	73.2	26.8
Developing countries	459	1 060	1 898	7.2	61.3	38.7
Latin America	168	287	463	1.7	74.9	25.1
North Africa	4	6	39	0.1	99.7	0.3
Other Africa	6	19	12	0.0	22.2	77.8
West Asia	20	22	109	0.4	76.7	23.3
South and East Asia	145	353	651	2.5	63.8	36.2
China	45	188	381	1.4	28.7	71.3
World	6 777[c]	13 913[c]	26 517[c]	100.0[c]	65.8[d]	34.2[d]

Notes
[a] SITC Rev 1: 541.7.
[b] SITC Rev 1: 512.86 + 541.1 + 541.3 + 541.4 + 541.5 + 541.61 + 541.62.
[c] Includes estimates for the USSR and East European countries.
[d] Excludes estimates for the USSR and East Europe.

Source: United Nations trade data; IMS (various issues); UNIDO estimates.

63

Table 3.2 *Leading exporters of pharmaceutical products, 1975–88*

Exporter	Value of exports (millions of current US$) 1975	1980	1988	Share in world 1988 (%)	Percentage share of exports (1988) Preparations in all pharmaceuticals	Medicinal chemicals in all pharmaceuticals
A World's major exporters			23 800	89.8	62.6[a]	37.4[a]
Germany, Fed. Rep. of	1 010	2 148	3 996	15.1	66.1	33.9
United States	725	1 772	3 541	13.4	47.5	52.5
Switzerland	860	1 629	3 172	12.0	67.3	32.7
United Kingdom	744	1 539	2 793	10.5	78.6	21.4
France	573	1 387	2 345	8.8	79.6	20.4
Italy	384	701	1 282	4.8	48.5	51.5
Netherlands	332	589	1 038	4.0	88.9	11.1
Belgium–Luxembourg	303	640	1 026	3.9	70.4	29.6
Denmark	145	315	839	3.2	78.0	22.0
Sweden	112	309	829	3.1	94.3	5.7
Hungary	214	322	739	2.8	62.3	37.7
Japan	124	296	721	2.7	22.3	77.7
Ireland	93	161	549	2.1	70.9	29.1
Spain	37	183	524	2.0	35.2	64.8
China	46	193	406	1.5	28.7	71.3

64

B Largest exporters among the developing countries and areas

			1 404	5.3	67.1[a]	32.9[a]
India	25	105	253	1.0	25.8	74.2
Bahamas	26	27	236	0.9	86.7	13.3
Yugoslavia	69	177	230	0.9	78.1	21.9
Singapore	72	148	164	0.6	67.2	32.8
Brazil	12	38	84	0.3	32.7	67.3
Taiwan	12	22	69	0.3	64.2	35.8
Korea, Republic of	8	20	68	0.3	52.5	47.5
Mexico	48	53	60	0.2	26.2	73.8
Jordan	4	9	50	0.2	100.0	0.0
Turkey	2	3	48	0.2	50.6	49.4
Egypt	3	5	32	0.1	99.8	0.2
Hong Kong	9	22	31	0.1	100.0	0.0
Argentina	24	37	23	0.1	45.9	54.1
Malaysia	5	9	18	0.1	97.2	2.8
Indonesia	12	12	17	0.1	54.4	45.6
Thailand	5	13	11	0.0	61.4	38.6
Colombia	8	12	10	0.0	98.3	1.7

Source: United Nations trade tapes.

Note
[a] Percentages expressed as unweighted averages for countries shown.

65

price. Competition through product innovation requires sophisticated research capabilities and financial power. These prerequisites confer a natural competitive advantage on producers based in industrialized countries. Other characteristics such as the high cost of marketing and the unusual standards of purity required in the production of drugs reinforce the export dominance of the multinationals.

A more comprehensive picture of trade performance is obtained from Table 3.3, which shows net trade in pharmaceuticals. Here a conspicuous feature is the durability of trading patterns. Almost all the countries that were net exporters in the mid-1970s had managed to increase their trade surpluses by 1988, while an opposite trend occurred among the net importers. Such consistency implies that differences in the competitive abilities of countries have probably become more marked since the 1970s.

The majority of the world's largest net exporters are industrialized countries, but their dominance is not so great as when total exports are considered. In fact, China's net trade exceeds that of several industrialized countries, while the Bahamas[4] and Yugoslavia have trade surpluses of at least $100 million. The list of net importers is longer but more evenly divided between industrialized and developing countries. Japan is a huge net importer with a trade deficit (mainly in preparations) which almost matched the surplus of the United States or the United Kingdom in 1988. In general all countries are actively involved in world trade but only a few developing countries export in amounts which would indicate a competitive advantage.

Given the dominance of industrialized countries, predictions about changes in the global pattern of competition have focused mainly on companies in North America, Western Europe and Japan. Some analysts draw parallels between pharmaceuticals and other research-intensive industries. They predict significant gains in the Japanese industry and a corresponding decline in the competitive position of American and European producers. Other commentators view this as an unlikely scenario, at least in the foreseeable future. They stress the continuity of the industry, arguing that the nature of competition between major groups cannot accommodate large swings in the relative position of particular groups within a period of only five to ten years.

To the extent that export performance can be regarded as a reliable indicator of competitive trends, the data support the latter view. The United States' share in exports of the industrialized countries has

Table 3.3   *Countries with largest net trade in pharmaceuticals, 1975 and 1988*

Country	Value of net trade^a 1975 1988 (US$ millions)		Average annual rate of growth, 1975–88 exports imports (per cent)	
A Net exporters^b				
Switzerland	680 (396)	2 312 (1 622)	10.6	12.8
Germany, Fed. Rep. of	500 (421)	1 858 (1 266)	11.2	11.7
United Kingdom	517 (460)	1 407 (1 141)	10.7	15.0
France	225 (433)	735 (1 183)	11.4	12.5
United States	490 (268)	667 (1 398)	13.0	21.2
Hungary	127 (116)	461 (304)	10.0	9.4
Denmark	59 (52)	450 (389)	14.4	12.2
Sweden	-65 (-40)	281 (324)	16.7	9.1
Ireland	33 (-22)	254 (170)	14.6	13.1
Bahamas	21 (24)	221 (191)	18.5	8.5
China	41 (28)	201 (30)	18.2	32.4
Belgium–Luxembourg	-32 (-16)	142 (863)	9.8	7.8
Yugoslavia	-5 (18)	99 (172)	9.6	4.4
Netherlands	98 (-9)	39 (-38)	9.2	11.8

Country or area	Value of net imports^c 1975 1988 (US$ millions)		Average annual rate of growth, 1975–88 exports imports (per cent)	
B Net importers^d				
Japan	308 (172)	1 399 (1 005)	14.5	13.0
Italy	-56 (-20)	546 (222)	9.7	14.1
Saudi Arabia	63 (63)	503 (518)	35.5	17.3
Canada	120 (40)	491 (320)	7.3	10.7
Hong Kong	74 (58)	478 (816)	9.6	15.0
Egypt	27 (16)	413 (244)	20.4	23.1
Australia	75 (-26)	407 (255)	9.1	12.6
Taiwan	18 (15)	345 (208)	14.7	22.4
Austria	75 (-27)	299 (300)	12.7	11.8
Algeria	114 (111)	260 (530)	-14.5	6.5
Iraq	50 (24)	229 (221)	-4.7	12.4
Norway	58 (58)	203 (196)	13.4	10.7
Finland	65 (59)	177 (153)	18.2	9.9
South Africa	52 (-6)	172 (-1)	-2.3	8.4
Pakistan	23 (-1)	163 (99)	7.4	15.8
Poland	13 (11)	159 (117)	-6.2	12.6
Venezuela	45 (13)	154 (36)	5.4	9.9

Notes
^a Net imports carry a negative sign. Net exports of pharmaceutical preparations are given in parentheses.
^b Countries with net exports exceeding $10 million in 1988.
^c Net exports carry a negative sign. Net imports of pharmaceutical preparations are given in parentheses.
^d Countries with net imports exceeding $150 million in 1988.

Source: United Nations trade tapes.

remained around 14 per cent since the mid-1970s, while the European Community countries account for over 60 per cent of the total. Japan's share has risen only slightly, and it is still less than that of smaller European countries such as Belgium, Denmark, the Netherlands and Sweden.

Producers in industrialized countries continue to dominate most drug markets but there are several reasons to explain why firms in developing countries may eventually have a larger role to play. First, demand in some of these countries is growing especially fast and domestic producers (whether local or foreign-owned) should benefit. Second, governments are anxious to encourage the production of multisource drugs. Multinationals and other drug firms in industrialized countries will probably claim the bulk of this new market, but in the longer term producers in developing countries should also gain. Third, the number of multinationals has grown and they compete more intensely than in the past. This should make it easier for governments in developing countries to establish local production since they will be able to choose from a larger number of potential foreign collaborators. Finally, firms in a few developing countries have launched an export drive on the basis of experience gained in the home market. In particular India's pharmaceutical industry can serve as an example of success in directing activities towards foreign markets (see Chapter 7).

In conclusion, at least some developing countries can be cautiously optimistic about their ability to compete in markets for certain drugs. For the present, however, these countries will continue to depend heavily on imports and/or local production by subsidiaries of multinationals. The following section takes up this issue, discussing the role of FDI in both the developing and industrialized countries.

INTERNATIONALIZATION OF PRODUCTION FACILITIES

Trade and foreign investment are alternative ways of establishing an international presence, but in the pharmaceutical industry the preference has always been for the latter. In 1980 about 13 per cent of the world's pharmaceutical needs were met through imports, while the corresponding share of local production by foreign-owned companies was 27 per cent (Rigoni, Griffiths and Laing, 1985, p. 5). Furthermore the changes since that time have almost certainly favoured foreign production rather than trade.

The fact that FDI is generally preferred to exports is due to both economic and policy reasons. Foremost among the economic factors are the segmentation of the global market into national markets, the oligopolistic structure of most submarkets and the importance of local marketing. Segmentation results from wide differences in country preferences. Local medical traditions are frequently the reason. They affect not only the formulation of drugs but even the choice of whole therapeutic classes of medicines. Doctors also contribute to this uniqueness since their prescribing practices usually favour locally-produced pharmaceuticals.

The oligopolistic structure of many drug markets is examined in some detail in Chapter 5. One fact which emerges from that discussion is that the pharmaceutical industry consists of a large number of heterogeneous submarkets where firm concentration tends to be high. Such a market structure is conducive to internationalization as national oligopolies with stable domestic markets look elsewhere in order to expand sales.

The provincial nature of many marketing tasks also contributes to the preference for foreign operations rather than exports. Marketing is typically a local activity which must take into account country-specific characteristics such as tastes and preferences, local disease patterns and regulatory systems. There may be substantial ownership advantages as well which lead firms to establish local marketing affiliates in addition to subsidiary production, formulation or packaging.[5]

The effects of public policy reinforce the tendency for firms to invest abroad. First, the 'natural' pattern of market segmentation is accentuated through the use of regulatory policies; for example, the registration of a drug may require that certain tests or phases of production be carried out locally. Second, policies governing prices and methods of reimbursement often create artificial barriers to trade since they favour local producers. Third, import controls can be a strong motive for firms to set up local production facilities in order to 'move behind' the trade barriers.

The degree of the industry's international involvement is suggested by the data in Table 3.4. The share of foreign sales, which includes exports as well as sales by foreign subsidiaries, is indicated for a sample of 143 firms with headquarters in 13 industrialized countries and three developing countries. The degree of foreign involvement is quite varied among companies based in industrialized countries. Japanese firms are the most dependent on domestic demand, while companies headquartered in France, the Federal Republic of Germany, the United Kingdom and the United States have a more prominent international profile. Foreign markets account for the largest portion of total sales in the smaller European countries such as Belgium, Denmark, Hungary and Sweden. In these cases a relatively strong international presence is needed to support a fully developed pharmaceutical industry.

The usefulness of Table 3.4 is limited by the fact that data for exports are lumped together with sales of foreign subsidiaries and there is no way to determine which are the more important. The United States is one of the few countries that reports separate figures on pharmaceutical exports and sales of foreign subsidiaries. These data, which are summarized in Table 3.5, show that exports have been no more than 17 per cent of total foreign sales and their share was falling between 1970 and 1987, when the trend was reversed. The share in other industrialized countries is probably higher but exports would probably still account for less than half of all foreign sales.[6] Intra-firm trade rather than sales to distributors or final consumers make up the bulk of exports. The extent to which companies trade within their own organization is clear evidence that strong links exist between the parent company and its foreign subsidiaries.

An alternative way to compare the competing roles of FDI and exports is to picture these operations in terms of resource flows to foreign markets. One set of resource flows can be represented by the amounts of productive factors embodied in drug exports. Another resource flow can be expressed in terms of the movement of productive factors across national borders in order to support foreign production. The latter measure or concept is more closely associated with FDI.

The relative importance of the two flows is difficult to assess empirically but indirect evidence seems to confirm that foreign investment is substantial in comparison with its 'export' equivalent. For example, data for the United States reveal that in 1983 about 50 per cent of the work force in American-owned companies was employed abroad.[7] Even if the capital/labour ratio is considerably lower in foreign subsidiaries than in parent companies, the share of overseas capital in

*Table 3.4   Foreign sales of pharmaceuticals by selected companies,
            averages 1987/88, by country[a]*

Country	Share of foreign sales in total sales of pharmaceuticals (per cent)	Number of firms covered
Belgium	82.5	2
Denmark	85.2	4
Finland	29.6	5
France	60.1	16
Germany, Fed. Rep. of	62.7	15
Hungary	54.1	8
India	12.4	7
Italy	32.3	13
Japan	6.2	28
Republic of Korea	5.1	12
South Africa	1.4	1
Spain	3.0	2
Sweden	79.9	4
United Kingdom	72.6	5
United States	42.4	26
Yugoslavia	39.6	3

Note
[a] The country averages can usually not be taken to be representative, owing to limited coverage of firms.

Source: Scrip, *Pharmaceutical Company League Tables* (various issues).

the total for the American industry would still be enormous.[8] Data for Swiss companies imply an even larger role for foreign investment. In two of the three largest Swiss multinationals the share of the foreign workforce was over 70 per cent of the total in 1981, while the corresponding share in fixed assets was 55 and 70 per cent, respectively (Tucker, 1984, p. 47).

*Table 3.5   Pharmaceutical sales abroad and exports by United
States firms,[a] selected years (millions of current US$)*

Year	Sales abroad	Exports	
		Total[b]	Intra-firm trade[c]
1970	2 052	344 (16.8)	244 (70.9)
1975	4 571	754 (16.5)	529 (70.2)
1980	10 035	1 219 (12.1)	739 (60.6)
1985	10 173	1 557 (15.3)	858 (55.1)
1987	14 281	2 013 (14.1)	1 226 (60.9)
1988	16 881	2 696 (16.0)	1 928 (71.5)

Notes

[a] The data shown here cover PMA member firms.

[b] The figures in parentheses show the percentage share of exports in total sales abroad.

[c] The percentage share of intra-firm trade in total exports is shown in parentheses.

Source: Pharmaceutical Manufacturers' Association (PMA), *Annual Survey Report* (various issues).

Country-specific figures such as these are helpful in forming impressions but provide little basis for generalization. To gain a broader impression of the role of FDI, data on total investment (both foreign and domestic) are presented in Table 3.6. Drug producers in the EC are the source of more than two-fifths of the pharmaceutical industry's global investments and their share has been rising. In turn, FDI accounts for a sizeable portion of the Community's fixed capital formation. A rough estimate of the magnitude can be derived from the share of EC pharmaceutical production accounted for by local affiliates of foreign companies. At the beginning of the 1980s this figure was about one-fifth of total EC production (Burstall and Senior, 1985, pp. 47 and 97) and FDI is assumed to account for a similar portion of total investment.

According to Table 3.6, the United States accounts for nearly a third of all fixed capital formation. Foreign-owned subsidiaries operating in that country had about 10 per cent of the domestic market in

Table 3.6 Gross fixed capital formation in the pharmaceutical industry, selected countries and years

Country/ Country group	Distribution among DMEs[a] (per cent)				Average annual growth (per cent)		
	1975	1980	1985	1988[b]	1975–80	1980–85	1985–88
Developed market economies	100.0	100.0	100.0	100.0	3.7	4.4	3.1
EC[c]	42.3	41.9	43.9	45.2	3.5	5.4	4.4
Japan	20.2	19.7	18.4	18.1	3.2	3.0	2.6
United States	31.7	32.4	32.7	30.5	4.1	4.6	1.0
Others[d]	5.8	5.9	5.0	6.3	4.2	1.0	11.5
Eastern European countries[e]	...	...	...	...	5.7	4.2	13.1
Developing countries and areas[f]	...	...	...	...	4.9	9.3	8.3

Notes
[a] At 1980 prices.
[b] Estimate.
[c] Belgium, Denmark, France, Germany, Federal Republic of, Greece, Netherlands, Portugal, Spain, United Kingdom.
[d] Australia, Austria, Canada, Finland, Norway.
[e] Czechoslovakia, Hungary, Poland.
[f] Bangladesh, Chile, Colombia, Ecuador, El Salvador, Guatemala, Hong Kong, India, Indonesia, Malaysia, Panama, Philippines, Republic of Korea, Turkey, Venezuela, Yugoslavia.

Source: United Nations Statistical Office, national pharmaceutical manufacturers' associations and estimates by UNIDO.

1975 (Burstall, Dunning and Lake, 1981, p. 63) but their share had risen to more than 17 per cent by 1988 (Imsworld Publications Ltd, *World Drug Market Manual*, 1989, p. 11). Thus FDI in the United States has grown quickly; by the end of the 1980s it was probably about 20 per cent of the industry's total fixed capital formation.[9]

Japan is the only industrial country where fixed capital formation is sizeable. The country accounts for about a fifth of the world total, and the share of FDI, though small, is growing. Local subsidiaries of foreign companies had just over 5 per cent of the Japanese drug market in the 1970s but by 1989 the figure exceeded 22 per cent. At that time, 15 of the top 50 companies in Japan were foreign-owned and FDI should continue to grow.[10]

The main source of FDI is the United States. European and Japanese firms are of less importance, although this may change in the 1990s (see Box 3.2). The United States firms are particularly strong in the European Community where their affiliates supplied around three-fifths of all the drugs sold by foreign subsidiaries in the early 1980s (Burstall and Senior, 1985, p. 97). Swiss firms are the second largest investors. They supplied roughly a fifth of all FDI in the European Community during this period.

Fragmented data seem to indicate that the same sort of pattern still applies today. In 1989–90, for example, companies based in the United States announced plans, began construction or completed plants in

---

Box 3.2   Japan's foreign direct investment

Japanese firms will become a much more important source of FDI in the future. One reason is the industry's research successes; only recently has it begun to develop products which are likely to be marketable in other industrialized markets. Firms are also being forced to turn to overseas markets in order to offset the costs of product development. Other factors which have contributed to the rise of FDI are the stimulus of foreign companies in Japan and the drastic price cuts imposed on the Japanese market.

At the beginning of 1990 Japanese companies had a total of 135 offices, joint ventures or acquired companies outside Japan. Preferred locations for research are Germany, Scandinavia, the United Kingdom and the United States. Sites for bulk production are concentrated in Ireland, Portugal and Taiwan Province, while much of the production of finished goods is found in France, Germany, Italy, Spain and the United States.

France, Italy, the Netherlands, Norway and the United Kingdom. Meanwhile Swiss companies were investing in France, Ireland and Spain. Intra-European Community investments were also initiated by a German company in France, by French firms in the United Kingdom and by a British company in Italy (Scrip, *World Pharmaceutical News*, various issues).

The developing countries are far less popular as recipients of FDI than the industrialized countries. No comprehensive figures are available but in the mid-1980s the volume of total investment (FDI plus domestic investment) was probably about a third of that going to industrialized countries.[11] Foreign-owned firms are nevertheless of immense importance in developing countries, accounting for about two-thirds of all pharmaceuticals produced in these markets.

Data for the early 1980s show that foreign-owned companies accounted for less than half of the domestic production in Bangladesh, China, Egypt, India, the Republic of Korea, Thailand and Turkey. In Chile, Mexico, Pakistan, Peru, the Philippines and Uruguay, the foreign sector was responsible for 50–80 per cent of total pharmaceutical production. In other developing countries such as Brazil, Colombia, Ecuador, Indonesia, Nigeria, Saudi Arabia and Venezuela, the output of foreign-owned companies was more than 80 per cent of the total (Redwood, 1987, p.262).

Table 3.7 provides additional information on this subject for the latter half of the 1980s. All the countries shown in the table have relatively large domestic markets and most have a number of domestic producers. Foreign-owned firms are a minority in every case. Nevertheless these firms are much bigger than their domestic rivals and typically account for well over half the domestic market. The multinationals' involvement varies, however, depending on the size of the market in the host country, the domestic industry's sophistication and the government's policies (see Box 3.3).

In developing countries where the industry is just beginning to emerge foreign firms are the main suppliers of medicinal chemicals and finished drugs in bulk form. Local activities are usually limited to packaging and dosage formulation. Developing countries with at least a modest pharmaceutical industry tend to rely on foreign-owned subsidiaries and the main source of competition is imports. By this stage in the industry's development most governments have adopted poli-

Table 3.7   Estimates of multinational involvement in selected developing countries, various years

| Country | Year | Number of companies | | Market share of foreign-owned companies (per cent) |
		National	Foreign-owned[a]	
Argentina	1986	69	58	50+
Brazil	1985	350	120	84
Chile	1985	37	30	60+
Colombia	1985	...	...	82
India	1986	289[b]	31[b]	23[c]
Indonesia	1989	247	40	72
Pakistan	1989	150	32	70
Peru	1987	46	18	80[d]
Philippines	1987	...	52	60
Rep. of Korea	1987	311	35[e]	17[f]
Turkey	1988	97	17	37[g]
Uruguay	1986	47	23	58
Venezuela	1984	39	41	82

Notes
[a] The definition of foreign-owned companies differs across countries.
[b] Figures refer to the 'organized' sector only.
[c] Share of top 13 foreign-owned companies in 1984.
[d] UNIDO estimate.
[e] Number of domestic–foreign joint ventures.
[f] Share of the top 22 domestic–foreign joint ventures.
[g] Figure refers to 1984.

Source: Scrip *Yearbook* (various issues), and IMS (various issues).

cies which favour indigenous firms relative to imports and foreign-owned subsidiaries (see Chapter 6).

The relationship with foreign firms is seen in a different light in countries where indigenous producers have the capability to manufacture pharmaceutical preparations. Besides foreign-owned subsidiaries

---

Box 3.3   Foreign direct investment in the Chinese pharmaceutical industry

The foreign sector accounted for less than 10 per cent of domestic production in the late 1970s. In the 1980s, however, Chinese policy makers came to rely more heavily on foreign capital and expertise to modernize their pharmaceutical industry. The country's huge domestic market and its long experience in traditional medicines made it an especially attractive market for foreign investors. As a result the State Planning Commission's efforts to encourage FDI for the production of certain antibiotics and traditional Chinese medicines (both made from constituents originating in the country) were remarkably successful.

The pace of FDI decelerated significantly in 1989, although existing assets and foreign investments in the industry were maintained. Examples include the joint ventures of Ciba-Geigy and SmithKline Beecham. The latter's operation in Tianjin is reported to have been profitable since 1988, mainly on the basis of large sales of an anthelmintic in the domestic market. FDI recovered in 1990 and 1991. Foreign companies involved in new operations include several investors based in Hong Kong and ICI Pharmaceuticals of California.

---

and joint ventures, local firms frequently produce brand-name drugs under licence from multinationals. The provision of production equipment and pharmaceutical chemicals represents other forms of linkages between indigenous producers and multinationals.

When the multinationals' involvement in developing countries is compared with that in industrialized countries, the differences are even starker. Some of these facts are documented in the Statistical Appendix, Tables A.4 and A.5. These tables show the location of foreign subsidiaries and research centres for a sample of firms from Belgium, France, the Federal Republic of Germany, Japan, the Netherlands, Sweden, Switzerland, United Kingdom and United States. Foreign subsidiaries operating in industrialized countries are all engaged in production and/or distribution and frequently carry out research as well. The same is not true for developing countries, where only a few companies are reported to have research facilities.[12]

In conclusion, the pharmaceutical industry in any country, whether industrialized or developing, depends to a significant degree on the participation of foreign multinationals. The multinationals' motives for FDI differ sharply between the two groups of countries, however. Their willingness to engage in non-restrictive forms of technology transfer – for example, minority-owned joint ventures, the creation of training facilities or research centres – is very limited in the case of developing countries. Such firms are nevertheless an important conduit for innovative breakthroughs and the international transfer of pharmaceutical science. Their involvement is essential if a country aspires to a competitive position in international markets, an aspect which is examined below.

INNOVATION AND INTERNATIONAL COMPETITIVENESS

Competition in the pharmaceutical industry occurs mainly through research-based product development. Research productivity, or innovative capacity, is therefore a key determinant of a firm's competitive standing in international markets.

Table 3.8 relates expenditures on R&D to the value of pharmaceutical exports (a rough indicator of trading strength) for 17 industrialized countries. The country rankings for the two variables are similar and rank correlations are strongly positive. When both variables are expressed in per capita terms (an adjustment which corrects for country size), the correlation coefficient is about 0.5 and is statistically significant at the 10 per cent level of confidence. The correlation is even stronger when no adjustment is made for differences in country size.[13] In that case, the coefficient is as high as 0.8 and is statistically significant at any level.

A detailed comparison of the data in Table 3.8 shows very few instances where the export pattern departs from expectations based on R&D expenditures. The major exception is Japan. That country's position as one of the world leaders in research spending sharply contrasts with its modest export volume. Some analysts attribute this anomaly to the industry's strong domestic orientation, but the result can also be taken to signify that even with large research spending a great deal of time is still required to establish an international position.[14]

The positive relationship between national rankings for R&D and export performance should have a parallel at the microeconomic level. Table 3.9 presents data on 31 companies for which information was available. When the company rankings are compared, the share of research expenditures and the share of foreign sales do, indeed, appear to be positively related.[15] Such a result can be interpreted in ei-

Table 3.8   R&D expenditures in relation to exports in selected
countries, 1986

Country	R&D expenditure[a] (millions of current US$)	Value of exports[a] (millions of current US$)
United States	3 548 (1)	2 869 (2)
Japan	1 859 (2)	521 (11)
Germany, Fed. Rep. of	1 534 (3)	2 990 (1)
Switzerland	1 082 (4)	2 342 (3)
United Kingdom	682 (5)	2 037 (4)
France	598 (6)	1 862 (5)
Italy	430 (7)	1 033 (6)
Sweden	230 (8)	580 (10)
Netherlands	115 (9)	776 (8)
Belgium	92 (10)	826 (7)
Denmark	65 (11)	605 (9)
Canada	56 (12)	80 (16)
Hungary	48 (13)	471 (12)
Spain	47 (14)	316 (13)
Austria	32 (15)	301 (14)
Norway	14 (16)	29 (17)
Ireland	11 (17)	244 (15)

Note
[a] Rankings within the given sample are shown in parentheses.

Source: OECD, Science and Technology Indicators Data Bank, United Nations trade
tapes.

ther of two ways. Substantial research expenditures may provide the
basis for a firm's international success and would be reflected by a
relatively large share of sales in foreign markets. Alternatively the
positive correlation can mean that success in international markets
allows the firm to develop a relatively intensive research programme.
In reality the causation should work in both directions.

Table 3.9 *R&D expenditures and foreign sales, selected companies,[a] averages 1987/88 (millions of current US$)*

Country	Company	R&D expenditure[b]	Foreign sales[b]	Total pharmaceutical sales
Finland	Orion	18 (10.9)	80 (35.5)	226
France	Rhône-Poulenc	310 (15.9)	1 631 (80.2)	1 945
	Roussel Uclaf	144 (14.7)	644 (65.2)	987
	Sanofi	177 (19.3)	528 (56.8)	920
	Institut Mérieux	38 (13.6)	159 (57.0)	280
Germany, Fed. Rep. of	Boehringer Ingelheim	369 (19.6)	1 554 (82.5)	1 884
	Schering AG	217 (15.6)	1 093 (78.5)	1 391
	BASF	87 (11.7)	493 (66.5)	740
	Merck	79 (11.3)	443 (63.7)	695
	Schwarz	28 (12.9)	54 (24.6)	217
	Asla Pharma	27 (13.1)	54 (25.8)	208
Italy	Sigma Tau	36 (11.2)	9 (2.5)	322
	Bracco	17 (5.9)	204 (73.5)	277
	Fidia	49 (21.4)	11 (4.6)	227
	Lepetit	31 (14.5)	89 (41.1)	213

80

Country	Company			
Japan	Yamanouchi	122 (9.5)	121 (9.5)	1 289
	Banyu	40 (6.2)	7 (1.1)	644
	Tsumura & Co.	44 (9.3)	0 (0.0)	474
	Zeria Pharma	25 (7.7)	0 (0.0)	320
	Toyo Jozo	31 (9.8)	30 (11.1)	311
	Tarii	10 (4.5)	0 (0.0)	215
	Tokyo Tanake	20 (9.8)	6 (2.7)	203
Sweden	Astra	181 (19.3)	775 (82.6)	938
	Pharmacia AS	129 (16.9)	613 (80.4)	763
United Kingdom	Glaxo	476 (11.9)	3 479 (87.3)	3 978
	ICI	274 (14.0)	303 (14.5)	1 950
United States	Warner-Lambert	219 (9.1)	571 (23.8)	2 399
	J & J	362 (16.7)	1 252 (57.8)	2 166
	Squibb ER	239 (11.8)	909 (44.7)	2 028
	Rorer	92 (9.3)	455 (46.1)	985
	Searle GD	199 (22.4)	523 (58.5)	896

Notes
a Only companies with sales over 200 million US dollars are shown here.
b The share in total sales is shown in parentheses.

Source: Scrip, *Pharmaceutical Company League Tables* (various issues).

Closer inspection reveals certain exceptions to this pattern. Some firms are successful exporters without devoting large amounts to R&D. Their achievements depend not so much on innovative leadership but rather on their abilities to distribute out-of-patent branded drugs or generics in international markets. Other companies (mainly in Japan and the United States) are heavily engaged in research but do not have substantial foreign sales. Instead the domestic market is sufficiently large and varied to accommodate these relatively large research outlays. Despite these exceptions, the relationship between research spending and international involvement is positive (as expected) and moderately strong. Apparently a country which 'produces' a fair amount of innovation is in a strong position to compete in international markets, either through exports or by establishing production facilities abroad.

The fact that competitive strength in international markets rests heavily on research capabilities means that producers in developing countries are mainly bystanders in the race for international competitiveness. These countries conduct some R&D but it is mainly of an adaptive nature, intended to modify products and processes to meet local conditions. The opportunities for these producers to compete in international markets will be limited to generic products for the foreseeable future.

In conclusion, this examination of trade, foreign investment and innovative capabilities suggests three salient features of the international environment. First, patterns of international competitiveness are rather stable over time and this continuity is expected to persist in the foreseeable future: competitive advantage in world markets depends mainly on research capabilities which are expensive and take a great deal of time to develop. Second, to the extent that changes of competitive relationships are observable, producers in industrialized countries seem to be strengthening their position: international differences in competitive abilities may even become more pronounced in the future; FDI has intensified but continues to follow established geographical patterns. Third, the prospects of producers in developing countries will depend mainly on the growth of world demand for generics and their ability to be price-competitive in these markets. Health care policies and other forms of regulation in both industrialized and developing countries will be an important determinant of the rapidity with which these markets expand in the future.

NOTES

1. International differences in competitive abilities can be gauged in several ways. One popular method relies on the theory of comparative advantage and makes use of an empirical measure of 'revealed' comparative advantage. In the case of pharmaceuticals, competitive differences depend mainly on elements of technological leadership and product differentiation rather than comparative advantage and this approach was not attempted. An alternative is to express international competitiveness in terms of market shares (see Burstall and Senior, 1985, p. 105). This method serves to distinguish between the leading exporters of pharmaceuticals and other producers but it is less well suited to draw a comprehensive picture of the international pattern of competitiveness. The approach used here is a more eclectic one which depends on export performance, foreign investment and innovative capacity.

2. Commodity fine chemicals (for example, acetylsalicylic acid, paracetamol, penicillin, quinine and vitamin C) are manufactured in large quantities by medicinal chemical standards in a relatively small number of dedicated plants (frequently classified outside the pharmaceutical industry). All pharmaceutical manufacturers, ranging from multinationals down to small generic firms, buy these chemicals in the open market. Commodity fine chemicals are sold also for the manufacture of products other than human pharmaceutical preparations (for example, veterinary drugs, animal feed additives, vitamin-enriched food products and tonic waters); consequently their prices are more flexible to demand than those of patent-expired specialty chemicals which are produced in relatively small quantities in a large number of multipurpose plants scattered around the world. Research-based pharmaceutical manufacturers produce their need for specialty chemicals in house, whereas generic companies buy them in the open market after the expiry of the basic patent. These chemicals are exclusively used for the manufacture of pharmaceutical preparations.

3. Among the suppliers of moderate importance, China and Japan rely mainly on sales of medicinal chemicals.

4. The Bahamas is an anomaly. The favourable rates of taxation have attracted a number of foreign companies, including producers of pharmaceuticals such as Syntex. Almost all of the island's drug production is for export.

5. For a detailed outline of the role of ownership advantages in international production see Dunning (1979).

6. Figures for the United States may not be representative of the situation elsewhere since American companies have been particularly aggressive in establishing a foreign presence.

7. Figures are based on Grebner and Reinhard, 1983, p. 14 and refer only to members of the United States Pharmaceutical Manufacturers Association.

8. The capital/labour ratio in foreign subsidiaries tends to be lower than for the parent company since most of the overseas investment is for the production of pharmaceutical preparations which is generally less capital-intensive than pharmaceutical chemicals.

9. This estimate is based on the market share of foreign subsidiaries operating in

the country market and the assumption that capital/output ratios are the same for local and foreign-owned companies.

10. FDI going to other industrialized countries would be relatively small, since their share of fixed capital formation amounts to only 6 per cent of the total for the country group.

11. This estimate is based on Table 2.1 and the fact that investment/output ratios are slightly higher in the developing countries than in the industrialized ones.

12. Sandoz has the most extensive network of subsidiaries in developing countries. Other companies, like Merck, Hoechst and Bayer, have a large number of subsidiaries, though most are located in the bigger countries of Latin America and Asia.

13. When the rank correlation is calculated in terms of the absolute values for R&D and exports, differences in country size are ignored.

14. Minor discrepancies can be noted for Canada, Norway and Sweden, where exports are somewhat less than might be expected on the basis of research spending.

15. The Spearman rank correlation between the two share variables is 0.57 and is statistically significant at the 0.1 per cent level.

4. The Role of Research and Development

Research (drug discovery) and development (the testing and approval of drugs) is one of the most crucial stages in the production of pharmaceuticals. The nature of competition in the industry places a premium on research; healthy firms must have a number of promising drugs in their research pipeline if they hope to thrive in today's markets. However research is expensive and its costs are rising. When R&D is expressed as a share of total revenue, the pharmaceutical industry has emerged as one of the biggest spenders in the manufacturing sector, ranking alongside aerospace, electronics and electrical equipment, and chemicals. Some of the implications of these opposing trends are considered below.

TRENDS IN PRODUCT DEVELOPMENT AND RESEARCH SPENDING

Although product development receives an especially high priority, the number of novel compounds being introduced to the market slumped in the 1980s. Many drugs in use today were actually discovered more than 20 years ago. The industry's fear is that its profits will be eroded as patents on existing products expire. The outlook for many pharmaceutical firms depends, to a large extent, on the pace of product development and their ability to channel more resources into research.

Global Pattern of Product Development

The pace of product development was frantic until the mid-1970s. On average, the world's pharmaceutical firms introduced 83 NMEs per year in 1961–74. Later the annual number of new drug discoveries declined, but by the late 1980s it had stabilized at about 50 NMEs per year.

Table 4.1 summarizes these trends, showing the total number of NMEs brought to world markets and their country of origin. Western European firms have been the world's leaders in product innovation, followed by American producers. Together drug companies in these two regions accounted for nearly 80 per cent of all the NMEs launched during 1961–90, though their leadership is now being challenged by Japanese companies. Japanese researchers averaged nearly 13 NMEs per year during the 1980s compared with only eight per year in 1961–80. Eastern Europe is another source of new drugs but the number of NMEs coming from this region has fallen off sharply: only ten were introduced to world markets in the 1980s, compared with a total of 58 in the previous decade. Few countries outside these four blocs can claim to have discovered an NME.[1]

The true 'nationality' of a NME is an ambiguous concept. Many discoveries are made in laboratories that are owned and operated by foreign firms. The research programmes of various countries also differ in terms of their originality and this, too, alters the global picture. A number of new drugs, for example, are incremental improvements on existing products which were first developed in other countries.

Table 4.1   Country of discovery of NMEs marketed world-wide, 1961–90 (number and percentages of total)

Period	Total number of NMEs	US	W. Europe	Japan	E. Europe	Other countries
1961–70	844	201	509	80	49	5
1971–80	665	152	375	75	58	5
1981–90	506	117	243	126	10	10
Total	2 015	470	1 127	281	117	20
Percentage	100	23.3	55.9	13.9	5.8	1.0

Source: Adapted by UNIDO from Reis-Arndt (1987) and calculated from information in Scrip *World Pharmaceutical News* (various issues).

Table 4.2   Drugs under development by corporate nationality for the top hundred ranked firms in 1989

Country	Number of firms	Self-originated drugs under development	Percentage of total
United States	34	1 190	40.0
Japan	28	623	20.9
Germany, Fed. Rep. of	8	250	8.4
United Kingdom	7	185	6.2
France	6	183	6.1
Switzerland	3	168	5.6
Italy	4	88	3.0
Others	10	289	9.7

Source: UNIDO calculations from Scrip Yearbook 1990.

Table 4.2 provides information which takes into account these characteristics. American-owned laboratories in the United States and abroad are the leader, accounting for about 40 per cent of the self-originated drugs under development. This is nearly twice the share of Japanese-owned laboratories, which have the second largest number of self-originated drugs under development.

This picture of research success is subject to one additional qualification: figures are based on a simple count of NMEs but some product breakthroughs are much more significant than others. The notion of 'consensus NMEs' (which is defined as the number of NMEs introduced in at least six of the world's eleven major markets) can be used to take account of this fact. Table 4.3 shows that, according to this measure, firms based in the United States have a substantial lead. American-owned companies actually developed 42 per cent of all the consensus NMEs in 1970–83. The combined total for the six European countries shown in Table 4.3 is 47 per cent. In contrast, Japanese firms claimed only 4 per cent of consensus NMEs during this period. The low share implies that Japan's laboratories have probably concentrated on imitative rather than innovative research. There is evidence, however, that this practice is changing. Japanese companies are now using the latest pharmacological concepts for research in a number of therapeutic areas, and the Japanese share of consensus NMEs began to rise in the late 1980s (Grabowski, 1989).

Table 4.3   *Distribution of consensus NMEs by nationality of originating firm, 1970–83*

Country	Number	%
United States	71	41.7
Switzerland	22	12.9
Germany, Fed. Rep. of	17	10.0
United Kingdom	17	10.0
Sweden	12	7.1
Italy	8	4.7
Japan	7	4.1
France	4	2.4
Others	12	7.4
Total	170	100.0

Note: Consensus NMEs are defined as new drugs introduced in at least six of eleven major markets over the period 1970–83.

Source: Grabowski (1989).

Research Spending

Despite the slowdown in product development, the industry's financial commitment to research has steadily grown. Figure 4.1 shows the pattern of research spending (in 1980 United States dollars) among the leading countries. Expenditures in the United States rose moderately during the 1970s and then began to accelerate rapidly. By the end of the 1980s, American outlays for R&D were more than three times their level in 1971.[2] Annual spending in the Federal Republic of Germany and Japan had reached almost $1.5 billion, while the outlays

Figure 4.1   R&D expenditures by the pharmaceutical industry in selected countries, 1975–89 (constant 1980 US dollars)

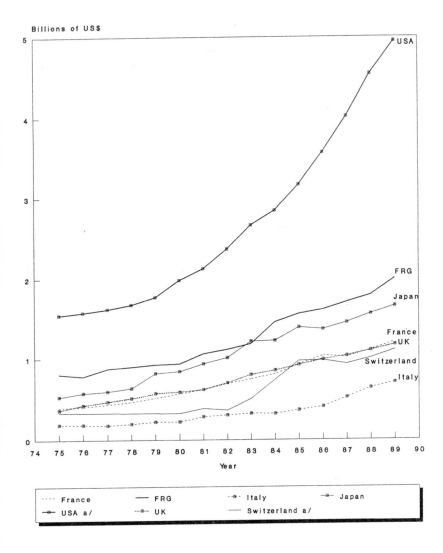

Note
a Including overseas R&D.

Source: Estimates derived from data supplied by national pharmaceutical producers' associations and the Office of Health Economics, London.

in other research-oriented countries were somewhat less – between $0.5 and $1 billion per year.[3]

When research spending is expressed as a share of total revenues, the amounts involved are clearly substantial. American companies, for example, were devoting roughly a fifth of all their revenues to research in 1989, up from 15 per cent in 1975.[4] Elsewhere the pattern is similar. Research expenditures in the United Kingdom were 5.7 per cent of gross output in 1970 but had risen to 13.3 per cent by 1988. Recent estimates for other European countries are even higher (Scrip, *World Pharmaceutical News*, 1433).

Sums such as these are large in comparison with the industry's turnover, but they understate the total amount spent on research because public funding is not considered. Government funding for bio-medical research exceeds company-financed expenditures in all but a few industrialized countries.[5] The United States government, for example, allocated $7.7 billion to biomedical research in 1988, a sum considerably greater than total company-financed spending on R&D. Public investments are an important source of knowledge for drug development. The results are available world-wide, though the countries where the knowledge is first produced probably have an advantage in using it.

As national spending on R&D has risen, so, too, has the research commitment of individual firms. The growth of company-financed R&D reflects an increase in the 'minimum threshold' for research spending. For example, the research costs incurred by a firm that hopes to market a NME world-wide are at present around $150 million and full-fledged activities can only be done with the firm's own staff for most of the development steps. These commitments require a minimum staff of 200, implying that operational costs are about $25 million per year. The minimum efficient operational budget for R&D in a firm with annual sales of around $170 million would therefore absorb 15 per cent of total revenues.

Table 4.4 gives a breakdown of operational research spending in one country, Denmark. The figures may not be representative of the pattern in countries where large, integrated producers dominate but they are useful indicators of circumstances in smaller countries where innovative firms are plentiful. The largest amounts of operational spending are for research management, the purchase of knowhow and pure research. Another third of the total goes to the development of

Table 4.4   *Operational spending on R&D of the Danish pharmaceutical industry, 1987*

Expenditure	Dkr million	%
Generic products	44	7
Products/process improvement	81	13
New products/processes	199	32
General accumulation of knowhow	22	4
Non-distributable costs[a]	280	45
Total	626	100

Note
[a] Includes costs of research management, information buying and pure research.

Source: *Facts 1989*, The Association of Danish Pharmaceutical Industry (1989).

new products and/or processes. Spending on operational research to improve existing products or processes is small – only 13 per cent.

Firms producing generic products are the only ones to have escaped the inexorable rise in research costs. Some of the reasons for this can be inferred from Table 4.4. Generic suppliers incur certain operational costs when acquiring products and knowhow and spend modest amounts to improve their existing products or processes. Their outlays for new products and processes are negligible, however, and non-distributable costs would be significantly lower than for research-based firms.

RESEARCH METHODS AND THEIR IMPLICATIONS

The costs of research are intimately related to the method of drug discovery. As these costs have risen, the industry's research leaders have been faced with a dilemma: competition forces them to come up with new products at a time when many are experiencing a decline in research productivity.

Traditional Methods of Drug Research

The process of drug discovery requires an extensive knowledge of the chemical makeup of compounds. Scientists had to master this field before they could begin to perfect systematic methods of synthesizing drugs. These efforts, which date back to the nineteenth century, led to the chemotherapeutic revolution in the 1930s. With traditional methods of drug research, scientists could create new compounds by changing the chemical makeup of known molecular structures. Afterwards they had to determine the physiological effects of the new compounds. The painstaking nature of these tests is illustrated by the search for a drug to guard against malaria. More than 230 000 chemical compounds were screened for anti-malarial activity between 1964 and 1974, though only about 30 were selected for clinical study in human volunteers (Heiffer, Davidson and Korte, 1984).

Traditional methods of drug research were tedious but also productive. A multitude of new and miraculous drugs began to flood the market soon after the end of the Second World War. They included antibiotics to fight bacteria, medicines to treat asthma, arthritis, cancer and heart disease, and contraceptives and vaccines.[6] Eventually it became clear that the key to success in the pharmaceutical industry was a strong and productive research programme.

The involved sequence of steps in the research process (outlined in Table 4.5) is one reason for the high costs. The types of analysis conducted during the pre-clinical stage are highly empirical. They include the initial screening process and a complex battery of tests (both in test tubes and in animals) to identify potential uses and to assess toxicity. Failure rates during the pre-clinical phase are extremely high: a rejection rate of 999 out of every 1000 test compounds is not unusual. Regrettably there is no way to predict the potential applications, if any, that each new compound might have. Thus the firm usually files for a patent before investigations have been started.

Promising products go on to clinical trials, where the first step is to conduct tests in healthy humans. If the drug is tolerated and produces the desired effects, it enters a second phase of clinical trials where it is given to patients who suffer from the ailment that the drug is expected to treat. Successful drugs are then tested in large-scale human studies involving sometimes as many as 25 000 patients. The large-scale tests help to determine ideal dosage and to refine safety and efficacy esti-

Table 4.5   Illustrative steps in drug development in the United States

	Total elapsed time
I Pre-clinical stage	
1. Synthetic chemicals are screened for potential use	
2. Pre-clinical studies are conducted in test tubes and animals	
Company files Investigational New Drug Application (IND) with FDA	2 years
II Clinical stage	
1. Clinical studies in healthy humans	
2. Clinical studies in patients	
3. Large clinical studies	
4. Company files new drug application (NDA) for review with FDA	8 years
III Drug approved for marketing	10 years

Source: Adapted from industry sources.

mates. Some of these trials may have to be repeated in the clinics of other countries where sales are planned. Ultimately about eight out of every ten drugs which began clinical trials are rejected. Roughly half of these failures result from the compound's instability. Other common reasons for rejection are lack of efficacy (23 per cent), undesirable side-effects (10 per cent) and toxicity (9 per cent).

Once a drug has successfully completed the development phase the company submits all the evidence compiled during clinical and preclinical trials to the regulatory agency, along with a request to approve the drug for sale. In the United States this process is known as a 'new drug application' (NDA) and is reviewed by the Food and Drug Administration (FDA). The review has typically taken up to three years but the FDA is now trying to reduce the time required and in special cases has cut the approval time to as little as ten months.[7]

Roughly ten years after tests begin, the successful drug will enter the market. About eight of those years will have been spent in development – two years for the pre-clinical tests, three more years to complete the first clinical tests in healthy humans and to carry out large-scale clinical tests with patients; in the United States a further three years are required for FDA approval. The entire patent life of a new drug is 17 years in the United States and 20 years in Europe, meaning that over half of this period is typically devoted to testing and regulatory review.

Research Costs and Productivity

Testing and regulatory review substantially reduce the period during which a marketable drug is protected by patent. The 'effective lifetime' of drug patents began to decline in the 1960s, just at the time when the flood of new drugs subsided.

The industry has long argued that government regulations are the real reason for both the decline in the effective life of patents and the rise in research costs. Many countries did, in fact, introduce more stringent regulations to control drug safety and efficacy in this period; firms were required to carry out much more extensive clinical tests and to provide supporting scientific documentation.

The tighter government controls imposed during the 1960s and 1970s almost certainly contributed to the industry's research dilemma but other factors were also relevant. First, several drug disasters were experienced, making it clear to consumers and government officials that existing regulations were inadequate. Second, the industry was probably slow to adjust its research practices to the new regulatory environment. Recent data show that rates of return on research began to rise in the 1980s and are now roughly equivalent to the cost of capital (see Chapter 6). Finally, the industry's traditional methods of drug development may no longer be as efficient as in the past and this, too, contributes to the fall in research productivity.

Figure 4.2 uses data for the United States to document both the rise in research costs and the decline in effective patent life. Real spending

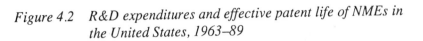

Figure 4.2   R&D expenditures and effective patent life of NMEs in the United States, 1963–89

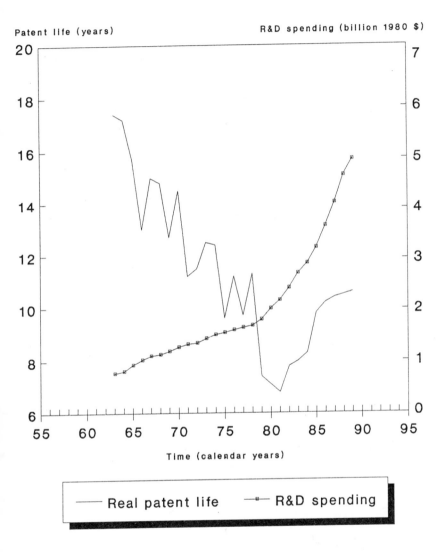

Sources: Estimates of patent life for 1963–81 were taken from Grabowski and Vernon (1983). Estimates for later years were done by UNIDO using average approval times reported by the United States Food and Drug Administration. Expenditure on R&D in current dollars were taken from Annual Survey Reports of the American Pharmaceutical Manufacturers Association and deflated to 1980 dollars by UNIDO.

on R&D doubled between 1963 and 1982. Meanwhile the effective patent life fell from 17.4 years to less than seven years. The increase in research costs persisted during the 1980s but the decline in effective patent life was reversed. Since 1982 the effective lifetime for drug patents has risen; it now exceeds ten years. The FDA cut year by year the time required for review of drug applications but the major cause behind the reversal was the introduction of more lenient government regulations.[8] One example is the Drug Price Competition and Patent Restoration Act passed in October 1984. Several other governments took similar steps to prolong the benefits provided to patent holders (see Box 4.1).

There is anecdotal evidence to suggest that the effects of a decline in the useful life of a patent may no longer be as crippling as in the past. Some holders of patents are voluntarily switching their protected products to the OTC market, apparently in the belief that the move will increase total revenues. The relevance of the true patent life, while still critical, is beginning to vary between firms. These inter-

---

Box 4.1   Policies to extend the marketable life of patented drugs

Japan introduced prolonged patent protection in January 1988. An extension up to five years could be requested if more than two years had lapsed between the registration and the approval of a NME. Innovative drugs are given protection from generic competition for six years after launch, irrespective of the patent situation. The patent extensions approved by the patent office range from two to five years, with an average of nearly four years.

In the United Kingdom, legislation passed in 1989 extended the effective patent life by curbing the issue of 'licences of right'. Previously, licensed copyists had been permitted the right to share the last four years of patent protection with the innovator.

The EC is also considering ways to extend the life of drug patents to compensate for time lost on safety testing before sale. The present life of a patent is 20 years, although companies argue that, because of testing, the effective life is only ten years. The EC is at present considering the use of Supplementary Protection Certificates. These would allow a drug up to 16 years of protection after its sale is approved. The European Patent Office seems to be in agreement with this approach but apparently wants the EC to revise its convention to permit any company (whether a pharmaceutical firm or not) to be permitted to apply for a patent extension under special circumstances.

firm variations result from a combination of changes in research methods and marketing strategies which are discussed in Chapter 7.

The steady rise in research costs poses a more serious problem for pharmaceutical companies than the issue of the patent lifetime because it directly affects research productivity. Table 4.6 provides an illustrative breakdown of research costs for a drug which is successfully marketed. The search for lead compounds absorbs between a fifth and a third of the research costs. Tests required to determine the potential benefits of these new compounds account for an even larger portion. Tests for pharmacology, toxicology and pharmacokinetics account for between 28 and 35 per cent of total research expenditures. Many of these results are required by government regulators but a number would have to be carried out anyway, since the firm has no

Table 4.6 Illustrative breakdown by activities and purpose of R&D costs of a typical NME (percentage)

	Activities	Estimated share in total research costs	Purpose
1.	Synthesis and extraction from natural substances	11–19	Search for lead compounds
2.	Biological screening	8–12	
3.	Animal pharmacology[a]	8–12	Verification of basic
4.	Toxicology and safety[a]	9–10	effects; determination of
5.	Metabolism and pharmacokinetics[a]	6–7	specific pharmacological
6.	Analysis research[b]	5–6	properties
7.	Clinical trials	16–28	Efficacy, safety
8.	Chemical process	10–12	Standard quality
9.	Pharmaceutical technology	7–10	Optimum dosage form
10.	Documentation for regulatory authorities	3–4	Registration

Notes
[a] Collection of data necessary for submission of IND application.
[b] Elaboration of testing methods must also be submitted with application.

Source: UNIDO, typical ranges compiled from annual reports of national manufacturers' associations.

way of knowing the characteristics of the new compounds it has identified. This fact undercuts, to some extent, the industry's claim that greater regulation has significantly boosted research costs.

One reason for the decline in productivity is that long research times force companies to speculate about future demand and plan their product development in the face of considerable uncertainty. Most have chosen to concentrate on the designing of drugs that are most likely to be commercial winners – in other words, medicines that meet an obvious demand and offer the prospect of high returns.[9] Table 4.7 shows how research activities are being focused on a few types of illnesses. Ten categories accounted for nearly 87 per cent of all research projects under way or launched in 1988 and more than half these were in one of five categories. The distribution of projects in the pre-clinical stage reflects the industry's assessment of markets around 1995, while those drugs which have been launched or are now in the clinical stage represent the industry's view of current market demand when research programmes were being planned around 1980.

The industry's guesses about future demand place a priority on anti-infectives, antineoplastics, cardiovasculars and biotechnology. The demand for anti-infectives is expected to rise because larger doses of established drugs and NMEs will be needed to combat higher levels of resistance. An increase in demand for cardiovascular drugs is forecast as the populations in major consuming countries grow older.[10] In the case of biotechnology, research is stimulated by the promise of new methods for the industrial manufacture of drugs. Human insulin, human growth hormones and other products derived from biotechnology have already been launched, while a NME for the treatment of cancer is nearing sales approval. New vaccines for malaria, rabies, cholera and other diseases common in developing countries may also appear as biotech-derived NMEs.

The uncertainties and rising costs of a large-scale research programme are common to many industries. However the reasons for the decline in research productivity in pharmaceuticals are more fundamental. Industry spokesmen now concede that traditional methods of drug research are becoming barren. The fact that many firms are at present in the process of changing their strategies for R&D is tacit acknowledgement that much of the problem is internal to the industry. This subject is considered in more detail below.

Table 4.7 *Top ten categories of R&D projects world-wide in 1988*

Research objective	Pre-clinical stage[a]		Clinical stage[b]		Drugs launch in 1988		Total	
	Number	%	Number	%	Number	%	Number	%
Anti-infectives	721	15.9	300	10.6	124	10.0	1 145	13.3
Cardiovasculars	509	11.3	395	14.0	170	13.8	1 074	12.5
Neurologicals	442	9.8	373	13.2	156	12.6	971	11.3
Antineoplastics	554	12.3	323	11.5	59	4.8	936	10.9
Biotechnology[c]	620	13.7	190	6.7	32	2.6	842	9.8
Formulations[c]	213	4.7	273	9.7	118	9.6	604	7.0
Blood and clotting products	317	7.0	133	4.7	69	5.6	519	6.0
Alimentary products	223	4.9	185	6.6	109	8.8	517	6.0
Musculo-skeletal products	215	4.8	153	5.4	94	7.6	462	5.4
Respiratory products	169	3.7	143	5.1	76	6.2	388	4.5
Top ten categories	3 983	87.5	2 468	87.6	1 007	81.6	7 458	86.7
Others	569	12.5	349	12.4	227	18.4	1 145	13.3
Total	4 522	100.0	2 817	100.0	1 234	100.0	8 603	100.0

Notes
[a] Includes chemical research and animal pharmacology experiments.
[b] Includes human pharmacology studies in medical research facilities.
[c] Refers to technological categories.

Source: Scrip, *Yearbook*, 1989, p. 51.

Alternatives to Traditional Research

Traditional methods of drug research are straightforward, though tedious and sometimes cruel (see Box 4.2). There is also growing dissatisfaction with these methods. One reason is that, over time, scientists and chemists have probably discovered most of the useful synthetic variations of familiar molecular structures. Another is the extremely high failure rate of new compounds which absorbs a large portion of all research spending.

These drawbacks have forced companies to search for other methods of drug discovery. Some have simply enlarged their scientific staffs, although this approach soon created its own problems: the increased size of research teams often meant that more than 200 scientists reported to the same R&D director. The resultant confusion inhibits discovery. More recently firms have begun to form smaller

---

Box 4.2　Animal tests and the alternatives

Animal experimentation declined between 1975 and 1985 but has stabilized since then. The reasons are mainly economic. First, a number of research establishments closed during the recession of the early 1980s. Second, the cost of animal testing has been rising – a monkey can cost more than $2000 to buy and another $100 per week to keep. Third, regulators have begun to insist that the animals be treated better. As a result fewer are needed to identify biological effects.

What are the alternatives to animals? Basic biological research accounts for 40 per cent of animal experiments and, in theory, can be conducted on cells rather than whole creatures. As scientists' knowledge of molecular biology grows, test tube (*in vitro*) screening will become a more reliable alternative to animals. However regulators have been slow to recognize these new methods: only two have been approved. Their argument rests mainly on a series of international trials involving a number of carcinogens. When the results of *in vitro* screening were compared with data from animals and people, there was only 60 per cent agreement. The problem remains that animal tests also give false results, rejecting safe chemicals or passing harmful ones. One example is Practolol, a heart drug which made people go blind although it had no known effect on any test animal.

Biotechnology firms are now at work to find better alternatives to conventional animal tests. Their goals are to develop better test tube techniques in order to reduce the number of animals used, to replace primates with smaller and less costly animals and to create better animal models of human disease.

research teams establishing separate units for biotechnology and various other areas.

Another strategy is known as 'rational' drug research. Instead of searching for new compounds and then trying to determine what they do, scientists apply the reverse logic. They first study what needs to be done and then attempt to design a compound which performs the task. Rational drug research has had some successes but the technical problems are great. Much depends on knowing the nature of the disease and understanding how it affects the body's chemistry. In most instances scientists have found that their knowledge of the disease is not sufficient for them to design efficient drugs to treat it.

Other attempts to raise productivity have involved a variety of piecemeal changes in research programmes. First, some of the more cautious firms in the United States and Europe began to concentrate on finding drugs for well-understood, long-term illnesses such as high blood pressure or arthritis. The idea behind this tactic is that most new drugs have benefited from fundamental discoveries in medicine.[11] Second, drug companies have begun to set up research facilities outside their home markets in the hopes of raising productivity. The main purpose is to exploit the scientific resources of other countries. This goal apparently outweighs the risk of duplicating work in different centres. A third piecemeal method is to redesign a firm's research programmes along academic lines. That tactic, too, has flaws. Academics who were brought in to guide research often proved to be unhappy managers and soon departed. Their involvement with the science community became too limited, while day-to-day managerial responsibilities absorbed excessive amounts of time. Nor does the practice of sponsoring research at universities guarantee a flow of new products. Universities are an abundant source of ideas, but there is little evidence that money spent on research outside a drug company is any more productive than that spent within the firm.

A general difficulty in devising new methods of drug research is that the range of scientific knowledge required today is much broader, involving not just chemistry but also biology and other sciences. In order to design new drugs, pharmaceutical firms must weld together a research team from independently-minded scientists and academics, all with different types of training and attitudes towards commercially-oriented research. This clash of philosophies can be aggravated by the secretive nature of pharmaceutical research. A firm's commercial success may depend on the progress of research in only one or two potential drugs. Much of the research which goes into testing a compound eventually proves to be of no value to the firm. Research directors may be reluctant to see any results, even negative ones, published, since the information could benefit competitors (not least by directing them away from unpromising lines of investigation). The scientists who carry out this work have a different attitude: they usually want to publish their results as soon as possible to obtain credit for their work.

By the mid-1980s all these efforts to accelerate research were overshadowed by the promise of biotechnology. Using their knowledge of deoxyribonucleic acids (DNA), scientists found that they were able to alter the structure of proteins. A desired gene was first isolated and then spliced into the genes of a common bacterium. As the bacterium multiplied, the gene was mass-produced. The early success of researchers led many in the pharmaceutical industry to predict that biotechnology would soon become the leading source for rare proteins which could then be used as drugs.

Table 4.8   Global sales of NMEs manufactured by advanced biotechnology in 1988 ($ million)

New drug	Sales
Alfa-interferon	80–140
Beta-interferon	5
Erythropoietin	15
Hepatitis B vaccine	70–300
Human growth hormone	150–250
Human insulin	150–450
Monoclonal antibody therapeutics	20–30
Tissue plasminogen activator	160–170
Total	650–1360

Source: Compiled from IMS Marketletter (various issues), and Scrip (various issues).

So far the results are disappointing. Table 4.8 shows that global sales of NMEs produced by advanced biotechnology are small. Total sales of all types of these NMEs were less than $1.4 billion in 1988. Enthusiasts underestimated the technical and commercial limitations of the new drugs. Contrary to expectations, human proteins give rise to as many side-effects as conventional drugs. They are often expensive to manufacture and must be given by injection. If proteins are swallowed as pills they are destroyed by enzymes in the gastrointestinal system and do not get into the bloodstream. An even greater problem is that proteins cannot be protected by patent since they can be found in nature and are not unique (see Box 4.3).

Pharmaceutical firms have not lost interest in biotechnology, although the field is likely to play a more modest role than was first envisaged. Many firms now regard it more as a tool for research than as a source of new products. Using the techniques of biotechnology, scientists are able to study the biochemistry of various diseases. Knowledge about the proteins which cause specific physiological responses is useful in setting the directions of R&D and therefore raising research productivity.

The question of how to organize a programme for drug development raises a host of issues which go far beyond the choice of research

---

Box 4.3   The legal complications of patenting biotechnology products

The ambiguous level of legal protection available for products created with the help of genetic engineering is a major problem. One difficulty is that the proteins developed by biotechnology firms are themselves found in nature. Because the results of the firm's research are not unique, there is much uncertainty about the degree of protection afforded by a patent. Another, more complicated issue involves second-generation products. Some of the molecules developed by biotechnology firms are identical to those found in humans. Others, however, have small structural modifications to the highly complex protein chain. The legal systems in some countries have concluded that the two versions are distinct and that the resultant products are different. Elsewhere courts have reached the opposite conclusion. In the latter case, the decision depended on whether the two products were 'biochemically equivalent' (that is, whether the molecules behave in the body in a similar way) and on the degree of similarity in the production processes. Many observers now believe that the legal questions relating to patents for biotechnology products are so complex that companies will increasingly rely on out-of-court agreements.

methods. These concern the organization of the firm, the types of alliances it may develop with other drug producers and the most suitable ways to carry out tasks of marketing distribution. The choice of a research approach is dependent on many other decisions and should be seen in this broader context. The following chapter carries this investigation to a more detailed level, looking at several of the microeconomic aspects of research, production and distribution.

NOTES

1. Other countries which have come up with NMEs include Argentina, Australia, Canada, China, India, Israel, Republic of Korea, Mexico, New Zealand, Portugal and Yugoslavia.
2. Around four-fifths of this sum was spent in the United States. The remainder was expended by American-owned firms carrying out research in overseas markets.
3. Such estimates must be interpreted with some caution. One reason is that various reporters – producer associations, national statistical offices and international organizations – use different definitions and concepts. A second source of discrepancy is that accountants treat R&D differently in different countries. Third, many companies conduct R&D in several countries, giving rise to ambiguities regarding the 'national origin' of expenditures.
4. The share of R&D expenditures in total world-wide sales (including sales by overseas subsidiaries) may provide a better estimate. For the United States this figure was 14.1 per cent in 1989.
5. The only developed market economies where private spending exceeds public spending are Ireland, Japan, Switzerland and the United Kingdom.
6. Contraceptives represent the first group of pharmaceuticals which are not used to prevent or to treat a disease.
7. For example, misoprostol, a prostaglandin derivative used in the treatment of stomach ulcers, was approved by the FDA in less than ten months in 1988.
8. In the United States, the new regulations grant a ten-year period of exclusivity for drugs approved between 1982 and late 1984, regardless of their patent status.
9. One regrettable outcome of focusing R&D on markets with commercial promise is the very limited amount of research carried out on diseases prevalent in developing countries.
10. Many of these pharmaceuticals must be taken regularly once the treatment begins.
11. For example, a better understanding of hypertension enabled companies to develop inhibitors which deactivate a specific enzyme responsible for raising blood pressure. Medical advances were also instrumental in the development of calcium blockers, which reduce the absorption of calcium into the blood-

vessel walls. Such an approach leaves the more laborious parts of product development to others (at present the Japanese). Finding new antibiotics means that firms must screen thousands of soil samples in their search for new antibiotics-producing organisms.

5. Competition and Cost Structure

Several of the industry's microeconomic features have been the focus of recurrent attention over the past three decades. One of the most crucial questions relates to the market structure and the degree of market power which exists in some parts of the industry. A second and closely related issue arises from the desire of governments to control prices. The priority attached to this regulatory activity has led to detailed analyses of the industry's cost structure. Thirdly, pharmaceutical companies differ in many ways but probably no microeconomic attribute is of greater significance than that of firm size. Differences in firm size are the crucial determinant of a firm's ability to compete efficiently and help to explain many of the changes occurring within the industry.

Each of these aspects is considered in this chapter. The first section looks at the pattern of firm concentration in domestic and international markets and examines the degree of competition in individual product markets. A large amount of cost data is reviewed in the second section and the cost structure in industrialized and developing countries is discussed. The third and concluding section of the chapter deals with the relationship between costs and firm size.

MARKET POWER AND PRODUCT COMPETITION

The question of whether some producers enjoy a substantial (or excessive) degree of market power and influence is an important one, not only for policy makers but for the industry and its customers. It cannot be answered with any degree of objectivity or precision, however. A more practical approach is to consider the degree to which market power and competition vary at different levels of aggregation (globally, nationally and at the product level) and whether the pattern has changed over time.[1]

*Table 5.1   World's 25 leading producers of drugs, revenues from
             drug sales and major products, 1988*

Pharmaceutical

1988	(1981)	Ranking company	Headquarters	$ m.
1	(3)	Merck and Co.	United States	4 984
2	(20)	Glaxo	United Kingdom	4 213
3	(1)	Hoechst	Germany, FR of	3 868
4	(2)	Bayer	Germany, FR of	3 628
5	(4)	Ciba-Geigy	Switzerland	3 466
6	(6)	American Home Products	United States	3 218
7	(8)	Sandoz	Switzerland	3 089
8	(18)	Takeda	Japan	3 076
9	(9)	Eli Lilly	United States	2 680
10	(17)	Abbott	United States	2 599
11	(5)	Pfizer	United States	2 539
12	(11)	Warner-Lambert	United States	2 509
13	(10)	Bristol-Myers	United States	2 509
14	(.)	Eastman Kodak	United States	2 500
15	(7)	Roche	Switzerland	2 365
16	(19)	J&J	United States	2 338
17	(12)	Upjohn	United States	2 234
18	(14)	Squibb ER	United States	2 213
19	(21)	Schering-Plough	United States	2 210
20	(16)	Rhône-Poulenc	France	2 016
21	(13)	SmithKline Beckman	United States	1 997
22	(24)	ICI	United Kingdom	1 921
23	(15)	Boehringer Ingelheim	Germany, FR of	1 866
24	(22)	Beecham	United Kingdom	1 846
25	(23)	American Cyanamid	United States	1 831
Totals				67 715

Note
[a] Estimates

Source: UNIDO, based on *Scrip Pharmaceutical Company League Tables (1989)*,
Scrip Yearbook 1990: *Scrip World Pharmaceutical News* (various issues).

Share of company-wide revenues (%)	Leading brand	Revenues from leading drug product	
sales		$ m.	Share in pharmaceutical sales (%)
83.9	Vasotec	1 000	20.1
100.0	Zantac	2 077	49.3
17.0	Claforan	376	9.7
16.1	Adalat	971	26.8
29.3	Voltaren	743	21.4
58.5	Inderal	356	11.1
45.4	Zaditen	428	13.9
64.8	Avan	328	10.7
65.8	Ceclor	605	22.6
52.6	Erythrocin	236	9.1
47.1	Feldene	616	24.3
64.2	Lopid	191	7.6
42.0	Questran	128	5.1
14.7	Omnipaque	450	18.0
40.6	Rocepin	476	20.1
26.0	Ortho-Novum	400	17.1
81.3	Xanax	395	17.7
85.6	Capoten	1 065	48.1
74.4	Proventil	159	7.2
18.5	Orudis	206	10.2
42.0	Tagamet	1 021	51.1
10.0	Tenormin	997	51.9
80.8	Persantin	337[a]	18.1
45.0	Amoxil	403	21.8
39.9	Minocin	201[a]	11.0
		14 165	20.9

Concentration in International and National Markets

When the pharmaceutical industry is pictured in global terms, a handful of multinationals are found to dominate. Table 5.1 documents the high degree of concentration which prevails in international markets. The top 25 companies reported sales of $67.7 billion in 1988, or 44 per cent of the world market for pharmaceutical preparations. Fourteen of these companies are based in the United States and have combined sales of more than $30 billion. Other company headquarters are scattered among Germany (3), Switzerland (3), United Kingdom (3), Japan (1) and France (1).[2]

The data in Table 5.1 also show that most multinationals are engaged in a range of activities other than pharmaceuticals. In fact most of the world's largest drug companies obtain the bulk of their revenue from the sale of non-pharmaceutical products. For example, Hoechst, Ciba-Geigy, Bayer and Rhône-Poulenc are primarily chemical firms with large pharmaceutical departments. Diversification of this type is sometimes regarded as a strategic advantage. That may be true in the sense that the very size of the company offers a degree of financial support which specialized competitors do not enjoy. However the fact that Merck, Glaxo and a number of other multinationals depend on pharmaceuticals for more than two-thirds of their total revenues suggests that this advantage, if it exists, is far from decisive.

The world's leading pharmaceutical companies are a rather select group. Entry into this club is very difficult, requiring large teams of researchers as well as a marketing network capable of distributing products on an international or even global scale. As a result the list of the world's largest firms has changed very little over the past several decades. There is, however, a great deal of change in the relative standing of the major firms. Most derive the bulk of their pharmaceutical revenues from the sale of a very few products. On average, 21 per cent of the pharmaceutical revenues earned by the top 25 firms come from the sales of a single product.[3] The introduction of only one or two new products can result in a rearrangement in the world rankings and market shares of leading multinationals. The commercial risks of the single-drug company are obvious but are often unavoidable, given the uncertainties of research and testing.

The picture of the industry is quite different when attention turns from the leading multinationals to the industry as a whole. Figure 5.1

shows the distribution by value of sales in a sample of more than 2200 firms operating throughout the world. Most companies are comparatively small. Those with sales of $5 to $25 million account for 43 per cent of the total number in industrialized countries and the preponderance of small producers is even greater in developing countries. In fact many companies in developing countries are so small (41 per cent report sales of less than $5 million) that they have no equivalent in industrialized countries. Their minuscule size means that issues like scale economies or research orientation play no part in day-to-day operations or, in many cases, long-term planning.

Questions relating to the 'domestic' structure of the industry are especially important to policy makers in developing countries but they are also of considerable interest for industrialized countries. One conventional way of looking at the industry's structure is in terms of the market share attributable to a predetermined number of firms. Several alternative expressions, known as concentration ratios, are reported in Table 5.2. Measured in terms of domestic sales, the share of the four largest firms in each national market is rather low in comparison with published figures for other industries.[4] In most countries, however, the ratio has risen since the late 1970s. The largest increases occurred in Italy, Venezuela, the Philippines, Peru, Argentina and Austria – in that order of listing – but the degree of concentration is not remarkably high in any of these countries.

Finally the concentration ratio will obviously rise when attention turns to the 25 and 50 largest firms in each market. However the upward trend is more pronounced in developing countries than in industrialized ones. Ignoring the possibility of competition through imports, the degree of oligopoly power is probably slightly greater in the former countries than in the latter.[5]

Competition in Product Markets

Estimates of concentration can be misleading since the markets for pharmaceuticals tend to be much more fragmented than those in other industries. Most manufactures are purchased by a rather broad group of users but medicines are sold in a number of self-contained submarkets made up of buyers with a specific disease or health problem. Industry-wide estimates therefore understate the extent to which a few companies dominate certain submarkets. In Mexico, for exam-

Figure 5.1   Size distribution of pharmaceutical firms in industrialized and developing countries, 1988

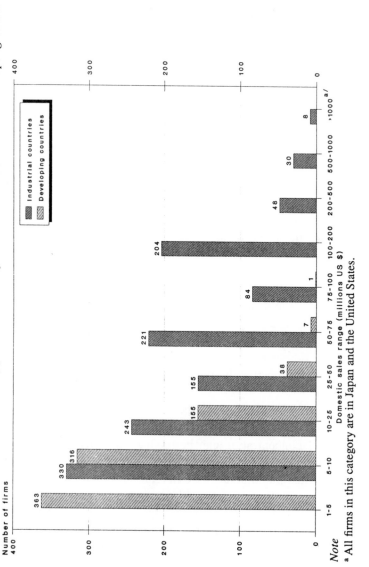

Note
a All firms in this category are in Japan and the United States.

Source: UNIDO calculations based on IMS, *World Drug Market Manual,* 1989.

Table 5.2   Concentration ratios in selected countries, 1978 and
             1988[a]

Country	Four-firm concentration ratios percentage change			25-firm ratio 1988	50-firm ratio 1988
	1978	1988	1978–88		
Developed market economies					
Australia	12	14	17	70	91
Austria	10	14	40	52	75
Belgium	12[b]	12	0	57	80
Canada	12[b]	15	25	58	83
Finland	33	39	18	86	97
France	8	11	38	46	67
Germany, Fed. Rep. of	9	9	0	38	57
Greece	11	12	9	57	77
Italy	7	14	100	50	69
Japan	16	17	6	60	82
Netherlands	15	19	27	60	83
New Zealand	13	15	15	69	...
Portugal	13	12	-8	55	79
South Africa	9	12	33	63	86
Spain	10[c]	10	0	45	67
United Kingdom	13	15	15	51	73
United States	13	14	8	57	76
Averages	13	15	15	57	78
Developing countries					
Argentina	13	19	46	63	86
Brazil	11	13	18	62	86
Chile	19[b]	21	11	73	97
Costa Rica	8[b]	9	13	49	68
Egypt	18[b]	16	-11	69	82
Indonesia	5[b]	7	40	58	82
Mexico	8	7	-13	59	80
Morocco	10	13	30	54	75
Pakistan	24	17	-29	67	83
Peru	9[b]	14	56	67	89
Philippines	10	17	70	61	83
Republic of Korea	...	17	...	...	...
Saudi Arabia	7	8	14	60	82
Turkey	17	17	0	70	87
Venezuela	7	12	71	67	91
Averages[d]	12	14	17	63	84

Notes
[a] The concentration ratio is defined as the share of domestic sales of the largest 4, 25 or 50 companies in total domestic sales of pharmaceutical preparations.
[b] 1980.
[c] 1979.
[d] Excluding the Republic of Korea.

Source: UNIDO, based on IMS (various issues).

ple, 64 per cent of the 300 largest-selling medicines were found to account for more than 40 per cent of their respective individual markets in the late 1970s. The situation is much the same in the European Community (EC): 50 or more brands exist in most product classes but the leading one will often have 20 to 25 per cent of the market and the top five may account for two-thirds or more (Burstall and Senior, 1985, p. 76).

The number of distinct product markets (referred to as therapeutic classes and subclasses in the literature) is great, reflecting the diversity of diseases and treatments which exist in any country. Some impression of the degree of competition can nevertheless be obtained from the data in Tables 5.3 and 5.4, which show the three best-selling brands in each of the three largest subclasses of selected countries. The range for the estimates is large but the figures generally confirm that a rather small number of drugs dominate in most cases.[6]

There are several reasons why competition in individual markets is not always vigorous. The most obvious is that some markets are dominated by a few, relatively efficient drugs which are patent-protected. In other instances the patents of the leading brands have expired but the products continue to be leaders. At first glance, this result seems peculiar; competitors would logically be expected to enter a profitable market once the patent of the dominant product has expired. The effects of patent expiry can be offset, however, by brand-name loyalty, control over a key input or policy decisions. The first of these factors is the source of much dispute. Large pharmaceutical firms spend huge sums on product promotion, a practice which has made them the target of considerable criticism. Effective promotion of branded drugs is clearly part of the mechanism by which returns to innovation are realized and the success of advertising will depend mainly on the therapeutic novelty and efficacy of the drug in question (Slatter, 1977). Nevertheless the hold of branded drugs is a powerful one (even in countries that do not grant patents). Some analysts suggest that brand-name loyalty may be a more effective method of guaranteeing high returns than the patent system itself (Lall, 1985b).

Evidence that drug companies go to remarkable lengths to promote their products adds credence to these arguments. In the United Kingdom, for example, the drug industry spent 80 times more than the country's national health service in 1988 to inform doctors about the drugs they offer.[7] Several companies have also established 'teaching

Table 5.3   Product competition in major therapeutic submarkets of industrialized countries, 1988, by product class

Countries	Total no. of products (year)	Total drug sales^a (US$ m.)	Share of largest-selling therapeutic subclasses in each national market			Share of 3 top products in the sales of the subclass		
			First	Second	Third	First	Second	Third
Austria	3 167 (85)	770[b]	4.2	2.7	2.6	45	22	33
Belgium	4 150 (88)	1 000[c]	5.5	4.2	3.3	44	75	64
Canada	17 000 (86)	2 900	5.5	4.3	2.2	31	74	54
Finland	3 606 (88)	438[c]	6.4	5.9	1.8	44	37	79
France	4 200 (86)	8 570[c]	7.0	4.4	2.0	33	38	66
Germany, Fed. Rep. of	8 792 (87)	9 700	5.4	2.4	1.6	41	46	44
Italy	5 309 (88)	8 000	6.8	5.3	4.3	43	65	32
Japan	15 933 (87)	26 600	11.0	4.7	2.7	30	42	76
Netherlands	6 139 (84)	1 037[b]	6.3	2.8	2.7	78	95	67
South Africa	3 623 (86)	500	6.6	3.0	2.6	42	64	41
Spain	6 000 (86)	2 670[d]	6.5	4.6	2.5	22	62	21
United Kingdom	2 090 (86)	4 350[c]	7.8	6.4	2.9	42	94	79
United States	19 000 (88)	30 900	5.4	5.2	3.1	87	42	39
Unweighted average	... ...	...	6.4	4.3	2.6	45	58	53

Notes

[a] Figures refer to sales through retail pharmacies and hospitals unless otherwise indicated.
[b] Sales through all outlets.
[c] Sales through retail pharmacies only.
[d] Sales through retail pharmacies and wholesalers only.

Source: UNIDO, based on IMS (1989) and Farmindustria (1989).

Table 5.4 *Product competition in major therapeutic submarkets of developing countries, 1988, by product class*

Countries/areas	Total no. of products (year)	Total drug sales[a] (US$ m.)	3 largest-selling therapeutic subclasses in share of national market			Share of 3 top products in the sales of the subclass		
			First	Second	Third	First	Second	Third
Argentina	3 500 (87)	1 100	5.9	3.9	2.6	26	52	31
Brazil	11 000 (87)	1 455	5.4	3.2	1.9	47	51	77
Chile	900 (87)	156	5.4	3.1	2.3	31	65	49
Colombia	...	370	7.0	3.5	3.0	31	41	57
Egypt	...	300	6.7	6.6	6.0	30	57	59
Indonesia	7 200[b] (79)	233	11.3	2.1	1.7	18	27	29
Mexico	7 000 (87)	944	5.8	4.9	1.5	41	72	45
Morocco	...	175[c]	6.3	5.2	3.6	55	52	65
Pakistan	9 700 (89)	391	6.7	5.7	3.0	66	31	27
Philippines	10 874 (88)	500[d]	7.4	4.0	3.0	32	36	57
Republic of Korea	12 358 (84)	1 533[d]	9.1	6.0	4.8	78	22	43
Saudi Arabia	3 500 (84)	314	5.7	5.1	4.1	66	51	41
Taiwan Province	...	420[e]	3.8	3.3	3.1	16	25	31
Turkey	1 862 (84)	482	8.4	5.5	5.4	56	38	49
Venezuela	2 848 (87)	180	4.9	3.3	3.2	35	47	61
Unweighted average	... ...	...	6.7	4.4	3.3	42	44	48

Notes

[a] Refers to sales through retail pharmacies.

[b] Number of presentations.

[c] Sales through retail pharmacies and wholesalers only.

[d] Sales through retail pharmacies and hospitals only.

[e] Sales through drugstores and private clinics only.

Source: UNIDO, based on IMS (1989).

centres' in attractive locations where doctors come to learn about medical matters and the company's drugs. The suspicion is that a portion of the industry's marketing budget is really intended to maintain brand-name loyalty rather than providing genuine information (see Chapter 7).

Control over a key input – for example, a medicinal chemical or active ingredient – can also mean that the original product retains a large market share long after its patent has expired. New competitors will require access to the input. The originator, however, may choose to sell the input to licensees, or else transfer the production technology if these methods are more profitable than sale of the drug itself. Such an outcome is most likely when the production technology is sophisticated and difficult to control (for example, the production of antibiotics by fermentation with good yield and consistently high quality). If competitors are unable to replicate the process exactly their ability to compete with the original product is severely limited.[8]

Finally policy decisions will affect the pattern of brand leadership. Various governments, for reasons of efficacy or cost, enforce programmes to ensure that only certain drugs are sold in the home market. The country may adopt the list of essential drugs recommended by the World Health Organization (WHO) or it may develop its own list. Some developing countries also distribute a large portion of all drugs through public channels and control the availability of drugs through the tender system they use for imports. The result is that only a few suppliers appear to dominate each market but, if government intervention and related measures – for example, price controls – are effective, the degree of market power should not be great.

The fact that the leading product is often an 'original' which is no longer protected by patent is changing the nature of competition in many markets. Traditionally the original market leader is expected to be replaced by a superior, research-based drug. That is still true when demand is growing rapidly, but in maturing markets such as antibiotics the amount of research is being cut back and the pace of innovation is slowing (see Box 5.1). As a result many battles for market share are fought not between two patented drugs but between an original and a copy. Factors like brand-name loyalty then assume greater importance. This helps to explain why firms are now willing to spend more to maintain the marketability of older drugs.

Box 5.1   The lifespan of products: evidence from Sweden

A recent study by H. Berlin and B. Jönsson shows an increase in the lifespan of products marketed in Sweden. Less than 60 per cent of the products launched in 1960–4 remained on the market for more than ten years, but the share has steadily increased over time (see Box table 5.1). Moreover the products of foreign manufacturers have enjoyed a longer lifespan than those of Swedish firms. The authors suggest that this is because foreigners only launch their better products in the Swedish market. To support this interpretation they note that the Swedish regulatory agency has initiated around a quarter of all withdrawals and its intervention has resulted in product modifications in other instances.

Box table 5.1   Product lifespan by origin of manufacturer, 1960–82 (in months)

Period	Number	More than 30 months %	More than 60 months %	More than 120 months %	Median
1960–4					
Swedish	596	89	77	55	129
Foreign	879	92	82	61	172
1965–9					
Swedish	327	95	80	59	146
Foreign	616	95	80	62	184
1970–4					
Swedish	320	96	85	71	...
Foreign	306	96	89	75	253[a]
1975–9					
Swedish	165	93	82	...	...
Foreign	306	95	89	...	263[a]

Note
[a] extrapolated
Source: Berlin, H. and B. Jönsson (1985).

The impressions which emerge from this examination of market power and competitive patterns are several. First at the international level the degree of market power is considerable and probably exceeds that in most other industries. The same does not apply at the national level. Most drug markets, however, are extremely fragmented and national data may be a poor indicator. The extent of market power and the limited degree of competition reappear when attention turns to the markets for individual drugs.

Second, the tendency for only a few drugs to dominate a particular product is widespread. Rival products, either patent-protected or copies, eventually appear but markets continue to be oligopolistic, marked only by changes in the leadership of firms. The nature of competition is changing, however, as firms become more judicious about the ways they spend their research funds. Research success and product innovation are still the main criteria for success in dynamic markets. Meanwhile research is being cut back in mature markets as firms rely instead on promotional efforts and characteristics such as brand-name loyalty.

Finally, the degree of market power and the extent of competition do not seem to differ significantly between the industrialized and developing countries. A few companies are world leaders and occupy prominent positions in the markets of both country groups. Nevertheless the implications for policy makers and consumers in the developing world are worrying. The domestic industry in these countries is relatively weak and consists almost exclusively of small firms which can pose no challenge to the multinationals. An added complication is that the regulatory system in these countries is often incomplete and sometimes inefficient (see Chapter 6). These circumstances mean that markets are relatively vulnerable to the possible abuse of market power.

THE PHARMACEUTICAL INDUSTRY'S COST STRUCTURE

During the 1960s and 1970s policy makers were mainly concerned with ways of regulating the oligopolistic behaviour of drug producers and focused most of their attention on measures of profits and rates of return. Today government officials are more interested in regulating prices rather than attempting to alter the market structure. This shift in emphasis has led to a much broader discussion of the cost structure.

The pharmaceutical industry is composed of firms of widely different sizes and forms of specialization. Such heterogeneity makes generalizations about the cost structure difficult, if not impossible. Table 5.5 sidesteps this issue for the moment, showing averages for a few very broadly defined cost categories. The distribution of costs has apparently changed very little in the developing countries, but the

Table 5.5 The average cost structure^a in developed market economies and developing countries, 1975 and late 1980s, as a percentage of gross output

Country group (number of countries)	1975				Late 1980s			
	Cost of inputs	Cost of labour	R&D costs	Other costs	Cost of inputs	Cost of labour	R&D costs	Other costs
Developing countries (26)	54.5	14.0	—	31.5	55.4	12.9	—	31.7
Developed market economies (12)	46.0	21.6	7.1	25.3	42.1	16.5	10.0	31.4

Note
^{a} Inputs represent the costs of materials and utilities and are defined as the difference between gross output and value added. The share of labour is represented by wages and salaries. Other costs include profit, administrative and selling costs and other expenditures. In some cases the data reported for gross output and value added included taxes or subsidies; whenever possible, the data have been adjusted to account for this fact.

Source: UNIDO, compiled from national statistical questionnaires.

same is not true for industrialized countries. In the latter case the share of non-labour inputs has been reduced and the developing countries' natural cost advantage of cheap labour has been eroded. These declines were offset by increases in the relative amounts spent on R&D and other inputs.[9]

Aggregate data of this type obscure any significant trends in individual countries or specific parts of the industry but they do suggest that the changes occurring since the mid-1970s were mainly in the industrialized countries. The absence of any significant research capability in the developing countries is also clear and this, too, would make for a much different type of cost structure. The pattern in industrialized and developing countries is examined in more detail below.

Cost Structure in Industrialized Countries

Table 5.6 provides a first look at the industry's cost structure in three of the countries which are among the leading producers. Manufacturing accounts for roughly two-fifths of all costs while marketing ab-

Table 5.6   The average cost structure in selected developed market economies, 1987 and 1988 (per cent of operating revenues)

Cost component	United States 1988	Switzerland 1987	Fed. Rep. of Germany 1988
Manufacture	35	40	39
Marketing	22	24	27
R and D	10	15	14
Administration	6	6	7
Other costs	6	5	6
Operating profit	21	10	7

Source: For the United States, company reports for eleven leading firms with calculations by UNIDO; for Switzerland, SSCI (1988); for the Federal Republic of Germany, UNIDO calculations based on PMAG, Pharma Daten 90 (1990).

sorbs between a fifth and a quarter. Research claims a smaller portion, though its share varies widely between the countries. Other cost components are of less significance, although the share of operating profits for firms in the United States is relatively high (21 per cent in 1988).

Manufacturing is the largest component in total costs. Because most producers use similar technologies, variations in manufacturing costs are not great so long as plants are of a comparable size. European plants tend to be of a smaller size than those in North America or Japan and the share of manufacturing costs probably differs accordingly (see Box 5.2).

Expenditures on research and marketing are more erratic sources of variation in total costs. A number of analysts have suggested that long-term changes in the shares of the latter two components are altering the industry's cost structure in a well-defined manner. Figure 5.2 illustrates this point, using data for a sample of large, research-oriented firms. The development which has attracted the most attention is the increasing share of marketing expenditures which, in the case of this sample, accounted for nearly a quarter of total spending in 1989. This shift was accompanied by smaller gains in the shares for R&D and operating profits and was offset by a large drop in the relative amount spent on manufacturing.

A widespread increase in the amounts firms allocate to distribution and promotion campaigns would cause concern among the industry's critics who fear that it represents a misuse of market power. The empirical evidence presented here is tenuous although anecdotal material offers more support.[10] Some of the larger drug companies, for example, have begun to change their basic approach to marketing.

Box 5.2   Plant size and rates of capacity utilization in Western Europe

The European industry's manufacturing capacity does not conform to the region's demand patterns or market size. In 1990 pharmaceutical formulation plants in the region were operating at very low rates of utilization – 50 to 60 per cent of capacity. This was far below the corresponding rates in North America. By the late 1990s large integrated firms will probably need only around ten strategically located manufacturing plants to supply local demand and to retain a global presence. Of these, no more than three to five should be in Europe. Such a development would imply a substantial reorganization of the European industry.

Figure 5.2   The changing structure of company costs in the pharmaceutical industry, 1973–89[a]

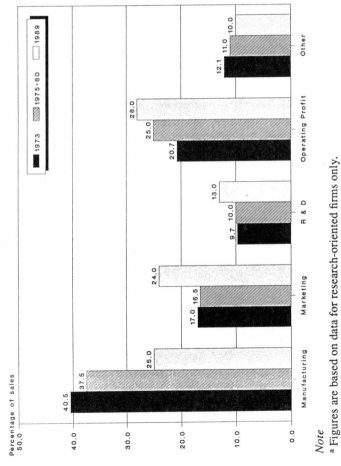

Note
[a] Figures are based on data for research-oriented firms only.

Sources: Based on Cooper, M. and A. Culyer (1973); OECD, 1975, p. 19; 1981, p. 31 and pp. 47–8; pharmaceutical company reports; UNIDO estimates.

Traditionally firms concentrated their marketing resources on a few national markets whenever they launched a new product. Now they try to launch a new product in all major markets simultaneously in order to maximize the revenues earned.[11] The adjustment requires a larger marketing force and heavy sales promotion. In fact the sales staff of the world's top pharmaceutical firms grew by 50 per cent in the period 1983–8. With such resources at their disposal, large firms are able to accomplish a world-wide launch in only three years, where they once required eight to ten years.[12]

This change in marketing tactics holds out some benefit for consumers since they would have access to the latest drugs with minimum delay. But the rising share of marketing costs may also reflect less admirable goals. One questionable motive – the desire to promote brand-name loyalty – has already been noted. Whatever the reasons, the reaction of various groups outside the industry seems to support the charge that marketing costs are rising disproportionately (see Box 5.3).

Box 5.3   Responses to the rise in marketing expenditures in the United States

Various government bodies and insurance companies argue that they are forced to absorb an unnecessarily large amount of the industry's growing promotional costs and question the need for such heavy expenditures. Consumer groups are also attempting to combat a perceived rise in marketing costs through joint buying of the most cost-effective medicines. More than 60 per cent of the drugs consumed in the United States are now bought through centralized operations which supply hospital chains. Doctors can then prescribe a drug only if it is on the supply list. The issue of who should bear the costs of promotion becomes even more contentious when companies are suspected of using these funds to preserve the market position of older drugs that face competition from new or cheaper rivals.

All this manoeuvring suggests that there are at least some grounds to suspect that marketing costs are rising too rapidly and could have several implications for the industry. First, the efforts of government officials and consumer groups to bypass the distribution systems of large firms could bring into question the industry's decision to increase outlays on marketing and promotion. Second, if the relative costs of marketing continue to rise, the industry will find it difficult to avoid direct forms of competition that have played only a minor role in the past. Drug firms that have developed large marketing staffs will be forced to compete more vigorously for a market share in the same way as producers of other consumer goods.

The possibility that marketing costs are unjustifiably high or are rising 'too fast' is a subjective judgement which cannot be resolved here. Whatever the case, one extraneous factor which is contributing to the rise in the marketing component is the growth of foreign direct investment. This fact can be illustrated with the help of Table 5.7, which shows the cost structure for domestic and foreign-owned firms operating in the Federal Republic of Germany. Foreign-owned firms tend to spend proportionately more on marketing (and significantly less on research) than their German-owned equivalents. Typically the main function of subsidiaries is to provide a local managerial base for the distribution of drugs which have been developed at research centres in the home country.

Table 5.7   Cost structure and ownership in the Federal Republic of Germany, 1988

Cost centre	Pharmaceutical firms		All firms
	German	*Foreign*	
Manufacture	37	46	39
Marketing	26	31	27
R & D	18	5	14
Administration	7	5	7
Other costs	7	4	6
Operating profit	5	9	7

Source: UNIDO, based on PMAG (1990).

In addition to production and marketing, research is the other significant component in the industry's cost structure. Because these activities are discussed elsewhere (see Chapters 3 and 4), they are not considered in any detail here. On average, research claims about 10 per cent of the pharmaceutical revenues in industrialized countries (see Table 5.5). The figure is much higher, however, for firms that are actively involved in the development of new drugs. The share of R&D in total costs declined slightly in the 1970s but in the next decade it increased by as much as 10 per cent per year. Real rates of return on research spending followed a similar pattern. They were around 6 per cent in the United States during the 1970s and rose to about 9 per cent in the 1980s (Grabowski and Vernon, 1990), implying that research activities were becoming more profitable. The costs are nevertheless huge, meaning that the bulk of this work is done by only a few large multinationals. The relationship between firm size and research spending is considered in the last section of this chapter, but before turning to that subject the cost structure in developing countries is examined.

Cost Structure in Developing Countries

The characteristics of drug producers operating in developing countries are quite different from those in the industrialized world and the cost structures vary accordingly. The scale of operation is fairly uniform, with most firms reporting sales of less than $10 million (see Figure 5.1). Product specialization occurs but not to the same degree as in industrialized countries. Nor do firms in developing countries have the same range of strategy options that producers in industrialized countries enjoy. Only a few possess the resources and personnel to mount even a small-scale research programme. The bulk of the industry consists of multinational subsidiaries whose main function is to market the drugs that have been developed by researchers at the company's headquarters.

A detailed breakdown of the cost structure of firms in developing countries is not possible. Instead, national averages for various cost components have been estimated from data obtained from industrial censuses, annual surveys or reports of industry associations. The results, which are found in the Statistical Appendix, Tables A.6 to A.16, provide the basis for a few, tentative generalizations.

First, the costs of manufacturing (which are mainly for the purchase of imported inputs) represent a much larger portion of the total than is true for firms operating in industrialized countries. Production-related expenses account for well over half of total costs in most of the countries shown in the Statistical Appendix. The distinction is to be expected since research activities are negligible. Expenditures on marketing are also of minor importance in several (though not all) developing countries, either because demand is limited or because the state controls the distribution of drugs.

Another significant feature is that the cost structure appears to vary with the country's level of development and the size of its home market. The average share of manufacturing costs tends to be largest (sometimes as much as three-quarters of the total) for firms operating in the smaller and less advanced countries. Spending patterns are different for pharmaceutical companies operating in big and comparatively sophisticated markets such as Argentina, Brazil or the Philippines. Manufacturing typically accounts for 50–60 per cent of total costs in these countries and the relative amounts spent on marketing are also somewhat higher (generally between 15 and 30 per cent of the total).

Such data are relatively detailed but refer to only a few countries and convey no impression about changes in structure over time. In order to supplement this information, additional estimates have been derived from other sources. Table 5.8 presents the results, showing expenditures on non-labour inputs (mainly raw materials and intermediate inputs), wages and other costs. The typical firm in a developing country spends, on average, more than half its gross revenues on inputs, another 13 per cent on labour, and 32 per cent for other purposes (mainly administrative costs, promotional expenditures and operating profits).

Variations in cost structure are great but most countries shown in Table 5.8 fall into one of two broad groups. In the first group are those where spending on inputs dominates, sometimes accounting for more than two-thirds of gross revenues. Reasons for inputs absorbing such a large portion of total spending differ within the group. In some instances, for example, Mauritius and United Republic of Tanzania, local production is either (very) small-scale or is mainly confined to the distribution of imported products. This is the situation in many developing countries with limited pharmaceutical capabilities (see Table 1.1) and the share of inputs in total spending will therefore be high.

There are other countries which have a significant pharmaceutical industry but still spend a relatively large amount on inputs. Examples are China, Ecuador, Egypt and India. China and India are important exporters of pharmaceutical chemicals, while Egypt produces medicinal chemicals on an industrial scale. Ecuador also meets a significant portion of its own needs for pharmaceutical preparations. The regula-

Table 5.8   *A breakdown of costs[a] in the pharmaceutical industry of developing countries, 1975 and the latest year, as a percentage of gross output*

Country/area	1975			Year	Latest year		
	Inputs	Labour[b]	Other costs		Inputs	Labour	Other costs
Argentina	39.2	17.5	43.3	1988	30.2	8.3	61.5
Bangladesh	56.2	10.9	32.9	1986	48.8	12.1	39.1
Bolivia	65.2[c]	16.8[c]	18.0[c]	1986	56.8	11.7	31.5
Chile	27.6[c]	14.6[c]	57.8[c]	1986	46.9	15.6	37.5
China	...	...	...	1986	72.9	3.8	23.3
Colombia	48.5	12.6	38.9	1988	56.0	8.6	35.4
Costa Rica	71.9	12.9	15.2	1988	66.9	9.3	23.8
Cyprus	...	...	...	1988	66.7	6.0	27.3
Ecuador	40.5	15.9	43.6	1988	78.1	18.0	3.9
Egypt	69.1	12.5	18.4	1986	73.7	13.9	12.4
El Salvador	50.9	11.8	37.3	1985	32.0	12.2	55.8
Guatemala	44.4	11.9	43.7	1988	46.9	9.4	43.7
Honduras	74.1	15.4	10.5	1988	67.3	16.6	16.1
Hong Kong	53.8	19.4	26.8	1987	59.9	16.4	23.7
India	68.3	10.9	20.8	1986	75.1	11.0	13.9
Indonesia	60.5	15.0	24.5	1986	65.6	12.9	21.5
Iran (Islamic Republic of)	20.6	12.4	67.0	1985	36.4	19.3	44.3
Jordan	...	...	...	1988	56.3	17.2	26.5
Kenya	70.2	13.7	16.1	1985	54.7	21.1	24.2
Madagascar	...	...	...	1986	64.8	10.5	24.7
Malaysia	52.0	17.4	30.6	1987	56.9	13.1	30.0
Mauritius	...	...	...	1988	77.6	9.3	13.1

Country		1975		Year			
Mexico	48.8	19.4	31.8	1986	49.3	18.2	32.5
Nepal	..	..	..	1987	63.9	15.3	20.8
Pakistan	57.4[d]	10.4[d]	32.2[d]	1986	55.0	10.7	34.3
Panama	51.3	13.7	35.0	1988	52.9	17.7	29.4
Peru	..	..	..	1987	61.3	10.9	27.8
Philippines	61.9	11.6	26.5	1987	64.7	11.6	23.7
Republic of Korea	50.4	10.7	38.9	1988	43.3	9.5	47.2
Somalia	..	..	..	1986	29.9	11.5	58.6
Sri Lanka	..	..	..	1988	55.1	14.1	30.8
Thailand	62.2	7.8	30.0	1986	61.6	10.0	28.4
Trinidad and Tobago	..	..	..	1987	67.3	19.5	13.2
Turkey	65.2[d]	15.2[d]	19.6[d]	1988	58.6	7.5	33.9
United Rep. of Tanzania	..	..	..	1985	79.8	9.8	10.4
Uruguay	..	..	..	1987	45.9	18.6	35.5
Venezuela	37.9	23.3	38.8	1989	47.4	15.8	36.8
Yugoslavia	67.7	9.4	22.9	1988	56.5	5.9	37.6
Zimbabwe	..	..	..	1986	51.0	16.3	32.7
Average[e]	54.5	14.0	31.5		55.4	12.9	31.7

129

Source: UNIDO, compiled from national statistical questionnaires.

Notes

[a] Inputs represent the cost of materials and utilities and are defined as the difference between gross output and value added. The share of labour is represented by wages and salaries. Other costs include profit, administrative and selling costs and other expenditures. In some cases the data reported for gross output and value added included taxes or subsidies; whenever possible, the data have been adjusted to account for this fact.

[b] The share of labour may also include social security expenditures if these are paid by employers.

[c] Data refer to 1976.

[d] Data refer to 1977.

[e] Excluding the countries for which data on 1975 are missing.

tory systems employed in these countries restrict profits and limit expenditures for administrative and promotional purposes, thereby increasing the relative amounts spent on inputs. In Ecuador prices are controlled at all levels. Regulators in several other countries in this group determine prices on the basis of costs which include expenditures for personnel, equipment, building materials and commercial and handling charges. China imposes an even stricter set of controls, since profits on sales are not returned to the manufacturer. A number of Indian companies are government-owned and many produce essential drugs at low prices, sometimes at a loss. The situation is similar in Egypt; the state-owned sector accounts for a large portion of the industry while private companies are controlled by a government organization.

A different type of cost structure applies to the second group of countries shown in Table 5.8. The drug producers included in this group spend proportionately less on inputs but allocate much larger relative sums to administrative and promotional expenditures and profits. The Argentine data provide a noteworthy example. The country has a pharmaceutical industry with innovative capabilities and produces both medicinal chemicals and pharmaceutical preparations. Its cost structure, however, is unusual: inputs are less than a third of total revenues while other components (excluding wages) account for nearly two-thirds. Argentine producers suffered a shortage of foreign currency in recent years which led to considerable delays in payments to foreign overseas suppliers and eventually resulted in a loss of commercial credits. The continuous devaluation of the austral against the dollar further reduced the drug companies' ability to pay for imports while pushing up their financial costs. These developments, together with high inflation and associated increases in social benefits and taxes, explain the large share of 'other costs' shown in Table 5.8. The Argentine experience represents an extreme case but producers in other developing countries have experienced similar problems which are reflected in the structure of cost.

In conclusion this survey reveals some marked differences in the cost structure. The spending pattern in industrialized countries depends mainly on the firm's own priorities but in developing countries public policy is the major determinant. In the former case, firms tend to devote a comparatively large portion of their funds either to research (15–20 per cent) or to marketing. Inputs, however, are the dominant cost component in a number of developing countries, while in others profits and expenditures on administration or promotion are more important. With this sort of cost structure, the prospects for the pharmaceutical industry in the developing countries during the next decade will depend mainly on improvements in existing production technologies, the acquisition of foreign technologies, the introduction of 'good manufacturing practices' (see Chapter 6) and the mastery of quality control.

Costs are also evolving along different lines in the two groups of countries. The real costs of research and marketing are steadily rising in the industrialized countries while the corresponding share for manufacturing and labour is falling. The structure continues to change as new products are developed, lines of specialization are altered and spending priorities are adjusted to shifts in policy and market conditions.

The same degree of adaptability is not to be found among producers in developing countries. The lack of significant research capabilities is one reason why these companies have few options. The small size of most firms also restricts their opportunities for specialization. Finally markets function very poorly in many of these countries and strong regulatory controls are essential. However frequent policy changes create a measure of uncertainty, which complicates the firms' efforts to control costs to introduce some form of specialization. As a result the activities of most locally-owned firms are confined to manufacturing, packaging or simply the distribution of drugs.

COST STRUCTURE AND SALES VOLUME

The industry's cost structure clearly depends on a number of variables, but one of the most important, scale economies, is yet to be considered. The concept, which relies on several simplifying assumptions, is a frequently used tool for analysis of cost behaviour (see Clark, 1985, Chapter 2). It leads to a stylized description of cost behaviour in which the long-run average cost curve is pictured as U-shaped. Such a curve implies that there will be some 'optimal size' for the plant. Firms of less than optimal size must incur relatively high unit costs but can make use of economies of scale if they expand.

Unit costs will eventually reach a minimum, however, after which they rise if the firm becomes larger still.

The application of this concept to the pharmaceutical industry can be informative but is by no means straightforward. A detailed breakdown of costs is needed for a number of firms of significantly different sizes. Furthermore, if these data are to be comparable across size classes, firms must be similar in terms of their research and marketing priorities, the production technologies they use and the products they produce. Such stringent criteria cannot be fully satisfied, though the data in Table 5.9 provide a reliable indication of the relationship between cost structure and sales volume of pharmaceutical firms with German ownership.

Of the six cost components identified in the table, three – administrative costs, operating profits and miscellaneous costs – are of secondary importance. Their combined share accounts for a substantial portion of the total among the smallest firms: 38 per cent in 1988. More important is that the share is inversely related to firm size. It falls for each successive size category.

The shares of the three remaining components account for larger portions of the total. Except category U2 in 1988, manufacturing claims more than two-fifths of operating revenues among firms in the three smallest size classes. Its share, however, falls to a little more than a third for the largest producers. Expenditures on research follow a different pattern. The smaller German firms devote less than 4 per cent of their revenues to research while in medium-size companies the figure is under 10 per cent. Large firms spend proportionally more on research and product innovation. In 1979, 16.3 per cent of the operating revenues in large firms went to R&D and by 1988 the share had risen to 19 per cent. These estimates – as well as the anecdotal evidence considered later in this chapter – suggest that the presence of scale economies in research activities is unlikely.

The figures for marketing present the most complicated picture. In the smallest firms these activities absorb around a fifth of all operating revenues. The share first grows with sales volume (reaching a third of the total among medium-sized firms) and then drops sharply among the largest companies in the sample.

In general, research, manufacturing and distribution are distinctly different operations and there is no reason to expect that the share of each would behave in the same way as a firm's sales volume in-

Table 5.9 Illustrative cost structures of German pharmaceutical firms in the Federal Republic of Germany, by size category,[a] 1979 and 1988 (percent of operating revenues)

Cost centre	1979						1988					
	U1	U2	U3	U4	U5	Total	U1	U2	U3	U4	U5	Total
Manufacture	44.7	42.0	42.4	38.1	38.2	38.9	40.3	35.4	41.1	35.9	35.6	35.9
Marketing	23.3	29.0	32.5	32.4	26.4	28.3	19.0	29.0	31.8	33.2	23.9	25.1
R&D	2.6	3.5	5.2	9.6	16.3	13.5	2.5	3.0	3.9	8.0	19.1	17.1
Administration	12.9	10.8	8.1	6.8	6.3	6.8	19.7	16.9	11.0	7.3	7.4	7.7
Other costs	6.9	6.2	5.5	5.0	6.9	6.5	9.0	7.3	5.9	7.5	8.1	8.0
Operating profit	9.5	8.4	6.3	8.1	5.9	6.1	9.5	8.4	6.3	8.1	5.9	6.1

Note

[a] Firm sizes are defined in terms of the annual average turnover in millions of DM, as follows: U1 – less than 7.5 DM; U2 – 7.6–15 DM; U3 15.1–45 DM; U4 – 45.1–150 DM; U5 – more than 150 DM.

Source: UNIDO, based on PMAG (1981 and 1990).

creases. Nor is each component of equal importance among firms of different size. Small companies, for example, may find that the costs associated with one particular component rise substantially as the

Table 5.10   *Ratio of R&D expenditures to total sales among pharmaceutical multinationals,[a] 1987 (percentage ratios)[b]*

Total sales (in US$ m.)	R&D expenditure as percentage of total sales
711–861	20.0
820–913	19.6
861–1 288	16.7
913–1 369	15.6
1 288–1 775	13.4
1 369–1 812	12.8
1 775–1 843	11.2
1 812–1 857	13.5
1 843–1 859	15.6
1 857–1 993	17.5
1 859–2 073	19.2
1 993–2 155	17.3
2 073–2 217	16.5
2 155–2 725	14.3
2 217–2 962	17.0
2 725–3 511	17.3
2 962–3 511[c]	19.0

Notes
[a] Of the world's 50 largest pharmaceutical firms, 19 report expenditures on R&D. Figures shown here are based on this sample.
[b] Ratios are calculated as three-company moving averages.
[c] Ratio includes only two companies.

Source: UNIDO, based on company reports and Scrip, *Pharmaceutical Company League Tables* (1988).

volume of sales grows. This relationship is maintained until some threshold is reached and only then does the share begin to fall.

In the case of manufacturing, the effects of firm size apply across a wide spectrum of companies. The same is probably not true for research or marketing, where changes in cost structure are most closely associated with producers having large sales volumes. It would be useful to explore each of these parts of the industry in more detail, but information is limited. Very few of the larger firms publish cost data on their marketing expenditures. A number, however, do provide information on research spending.

Table 5.10 makes use of this information, relating expenditures on R&D to total sales. The share of R&D declines as sales rise, reaching a minimum (around 11 per cent) when sales are approximately $1.8 billion.[13] Research expenditures account for a much larger proportion of revenues among those multinationals with sales of under $1 billion. Thus the world's larger pharmaceutical firms are able to reduce the burden of research costs provided that they have a large and efficient distribution system which operates in a number of markets.

The impression which emerges from this investigation is that sales volume is an important cost determinant in the manufacture and distribution of medicines. It is of less significance in the case of research. A large size is not a prerequisite for success, however. Even in the industrialized countries there are a number of firms which operate with low overheads, national or regional sales forces and which conduct only limited development activities. The smaller firms thrive because they have identified market niches which their larger competitors have neglected or cannot satisfy. Such diversity multiplies the number of issues to be addressed by policy makers and corporate strategists. Both these topics are discussed in the following chapters.

NOTES

1. Economists usually choose to discuss market power in terms of contestable markets. The notion of contestability stipulates that potential entrants need not incur any costs which are specific to entry or exit. The investments in plant and other assets that are required for entry would be recoverable on exit (after allowance for depreciation). However a new pharmaceutical firm's outlays for research and distribution are great and a portion would not be recoverable.

The idea of contestable markets would call for microeconomic policies that increase the contestability of drug markets.

2. The fact that only one Japanese firm appears in Table 5.1 is somewhat misleading. When pharmaceutical firms are ranked in terms of growth of sales, five of the ten fastest-growing are Japanese.

3. SmithKline Beecham (SKB), for example, obtains more than half of its pharmaceutical revenues from one drug, Tagamet. ICI, Glaxo and Squibb are other companies which depend heavily on the sale of a single drug.

4. The general tendency is for capital-intensive and high-technology industries such as vehicles, metal manufacture or electrical engineering to be highly concentrated, while labour-intensive industries like leather, footwear or clothing are not so highly concentrated.

5. Imports, of course, are an important source of supply for many developing countries and any conclusions based purely on domestic measures of market structure may be misleading.

6. The sales volume in many of these individual markets is huge. In the United States, for example, the three leading brands in the largest therapeutic subcategory report combined sales of nearly $1.5 billion. Even in developing countries some submarkets are large and often lucrative for the suppliers of the best-selling drugs.

7. Apparently the tactic is an effective one. According to one recent survey, six out of ten British physicians prescribe a drug solely on the basis of what a drug salesman has told them.

8. Most industrialized countries require that the production premises of reproductive manufacturers of medicinal chemicals be inspected and approved by their own health authorities before imports are allowed. Thus it may be extremely difficult for new competitors to supply products which make use of intermediate inputs using sophisticated technologies. Most firms in industrialized countries obtain their inputs from licensees or affiliates of the originator if not from the original company itself. The same is not true in developing countries, where inspection of production facilities is rare. Instead the latter firms usually obtain most of their inputs through international trade.

9. These shifts were actually very broadly based trends. The share of R&D rose in each of the 12 industrialized countries for which data were available. The pattern was much the same for other unspecified expenditures; these rose in 11 of the 12 countries considered here.

10. The estimates in Figure 5.2 are subject to three qualifications. First, the composition of firms in the sample changes over time period and the results may not be representative of trends in most large drug companies. Second, firms are not consistent in the way they define their costs. Where possible, company data were standardized to account for these discrepancies but in many cases sufficient information was not available to carry out these adjustments. Third, some of the larger drug producers have now ventured into generic markets. Such a move alters the cost structure but could be independent of a more broadly based shift in the industry-wide composition of costs.

11. This is one reason for a number of firms having sought to merge or acquire competitors. Their goal is to become big enough to sell new drugs in large volumes as quickly as possible. Such tactics, which go beyond the field of microeconomics, are among the strategies discussed in Chapter 7.

12. Glaxo was one of the first firms to break with this precedent by undertaking a world-wide launch rather than focusing on just a few national markets. The firm spent lavishly to promote its new anti-ulcer pill. Within only a few years the drug became Glaxo's best-selling product, even though it was probably only a slight improvement over its closest competitor. Contrary to the usual practice, Glaxo did not wait for all the drug's possible uses to be approved. Sales of Glaxo's anti-ulcer pill reached $2.4 billion in 1989, double that of its closest competitor.

13. Other calculations (not shown here) seem to confirm this pattern. When the ratio of R&D expenditures to profits is related to total sales, a similar pattern is observed. Again the total value of sales where a minimum is realized is around $1.8 billion.

6.   Policies and Issues

The pharmaceutical industry in most industrialized countries enjoyed a good working relationship with government during its early years. This congenial atmosphere reflected the view that drug firms were making a substantial contribution to the public's health and, through exports and employment, to the country's wealth. Policy makers were generally inclined to accept the industry's arguments that increases in prices and profits were essential if the research to develop cures for major diseases were to continue.

The United States was one of the first countries where government-industry relationships were marred by a major confrontation. Questions about the industry's monopolistic practice and the social benefits of its research led to a congressional investigation. The legislation which emerged from this investigation set new standards for pharmaceutical products (Box 6.1). The environment in which drug companies operated soon became even more hostile following the highly publicized thalidomide disaster.[1] Doctors grew more cautious in their prescribing habits and consumers became suspicious of the medications they received.

Pharmaceutical companies had traditionally performed only voluntary checks on their products. As concern about the physiological effects of pharmaceutical products mounted, the need for a more formal and rigorous set of procedures became obvious. The pharmaceutical industry soon became a prime target for regulation. The new regulatory systems increased the costs of product development but the lack of such controls would have imposed even larger costs on society.

The safeguards which exist today are much more comprehensive than those of the 1960s but they are far from perfect.[2] Drug disasters can still occur: it is simply impossible to predict from clinical trials the full range of reactions to a product in the real world. In a few instances the regulatory system has been abused, when government reviewers have accepted bribes or drug companies have falsified test

---

Box 6.1   Setting standards for drug efficiency

The regulations which were in force in the 1950s and 1960s seem lax by today's standards. Scientists were not always expected to carry out extensive safety investigations involving tests for mutagenicity, toxicity or long-term carcinogenicity. Nor were governments especially concerned about efficiency so long as they thought the product was safe. An example is papaverine, a cerebral vasodilator. The drug continued to be sold under more than 40 brand names long after human clinical trials proved that it was inefficient.

One outcome of the congressional investigation in the United States was that drug companies, for the first time, were required to prove the effectiveness of an NME before it could be marketed. A new form of application known as the Investigative New Drug (IND) procedure was also required before human experiments could be conducted.

---

results. Even worse is the growing number of drug forgeries. Pirate drugs, which may be medicines of greatly reduced strength or even harmful counterfeits, threaten to undermine the consumer's confidence. Such occurrences do not necessarily reveal glaring weaknesses or gaps in regulatory systems but they have amplified the calls for even stronger controls on the industry.[3]

Safety and efficiency are only two of the issues which concern government officials. Pharmaceuticals is a particularly sensitive industry in terms of the debate it arouses and the regulations it invites. As a result the policy landscape of rich and poor countries is cluttered with all sorts of controls, testing systems and other forms of intervention. This chapter begins with a survey of some of the policy tools currently in use. Later sections discuss the policy-related issues which figure prominently in industrialized and developing countries.

THE POLICY FRAMEWORK[4]

Table 6.1 focuses attention on price controls in a number of industrialized and developing countries. Price controls are the most prevalent form of intervention. The low incomes and incomplete systems of public health care explain why prices receive such a high priority in developing countries. Officials in industrialized countries have also become more concerned with this issue as their populations grow older and push up the costs of public health care.

Almost all governments regulate product prices, though a few choose to limit profits or to influence prices through more indirect means. Both governments and insurers are usually involved in the price-setting exercise in industrialized countries. In developing countries, however, a majority of the population is not covered by health insurance and price regulations are almost exclusively in the hands of public officials.

Systematic differences in the extent of price regulation can be noted. Industrialized countries are fairly equally divided between those with elaborate systems of controls and those that allow firms some freedom to determine their own prices. A majority of developing countries have rather extensive systems of price control. Some of the reasons – for example, the prominent role of foreign subsidiaries and the limited purchasing power of consumers – are explored later in this chapter.

The actual methods of price control differ. Most countries choose to regulate both retail and wholesale prices. Price-fixing procedures also vary, depending on whether the drug is a new entrant to the domestic market or an established product for which a price increase is requested. Finally some governments merely stipulate the maximum price; others set the actual price and frequently apply different procedures for locally-produced products and imports.

The contrasts between industrialized and developing countries are sharpest in the case of patents. Almost all industrialized countries grant patents on both products and processes – typically for a period of 20 years. The practice in developing countries is more varied. Only 45 per cent of the studied countries grant product patents and these are usually valid for a shorter period of time than in industrialized countries. Patents on production processes are more common in developing countries although, again, the period of validity is comparatively brief.

Manufacturing guidelines are a third important feature of national policy described in Table 6.1. Known in the industry as 'good manufacturing practices' (GMP), the guidelines are an application of the principles of quality assurance which initially related to the manufacture, processing and packing of pharmaceutical preparations; gradually the scope has been widened to include medicinal chemicals as

Table 6.1   Selected policy features in industrialized and developing countries

A   Industrialized countries

Country	Price controls	Product	Patent[a]	Process	Manufacturing guidelines	Research/policy support
Australia	Limited		16–20 years		WHO	EF, PP1
Austria	Substantial		20 years		WHO	EF, PP1
Belgium	Substantial		20 years		WHO	EF, PP1
Canada	Limited		20 years		Local	NP
Denmark	Limited		20 years		PIC, WHO	EF, PP1
Finland	Substantial	Beginning in 1995		20 years	WHO	...
France	Substantial[b]		20 years		WHO	EF, PP1
Germany	Limited		20 years		WHO	EF, PP1
Greece	Substantial		15–20 years		PIC	...
Hungary	Substantial	No	20 years		PIC	EF, PP1
Ireland	Limited		16 years		UK	EF, PP1
Israel	Substantial		20 years		WHO	EF, PP1
Italy	Substantial		20 years		WHO	EF, PP1
Japan	Limited		15–20 years		WHO	EF, PP1
Netherlands	Limited		20 years		Local	PP2
New Zealand	Limited		16 years		WHO	EF
Norway	Substantial	No		20 years	PIC	EF
Portugal	Substantial	Beginning in 1992		15 years	PIC	PP2, NP[c]
Spain	Substantial	Imminent		20 years	PIC	EF, NP[c]

Sweden	Substantial	20 years	WHO, PIC	EF, PP1
Switzerland	Limited	20 years	PIC	...
United Kingdom	Limited[d]	20 years	Local, WHO, PIC	EF, PP1
United States	Limited	17 years[e]	FDA regulations	EF, PP1

B   Developing countries/areas

Algeria	Substantial	No	WHO	EF for traditional medicines
Argentina	Limited	15 years[e]	No formal requirements	PP1
Bangladesh	Limited	16 years	Limited	...
Bolivia	Limited	5–15 years[e]	Local	PP2
Brazil	Limited	No	Local	NP
Chile	No	15–20 years[e]	Local, based on WHO	...
China	Substantial	No	None	EF for traditional medicines
Colombia	Substantial	10 years[e]	Local	...
Costa Rica	Substantial	1 year[e]	None	...
Egypt	Substantial	10 years	Limited	EF for traditional medicines
Ghana	Substantial	No	None	...
Hong Kong	No	Yes	Informal legal provision	...
India	Substantial	7 years	USA, FDA	EF, PP1
Indonesia	Limited	No	...	...

143

B *Developing countries/areas continued*

Country	Price controls	Product	Patent[a]	Process	Manufacturing guidelines	Research/policy support
Iran (Islam, Rep. of)	Substantial	No		20 years		EF, PP1
Jordan	Substantial	No		16 years	None	...
Kenya	No		Yes		None	EF for traditional medicines
Malaysia	No		15 years[e]		Voluntary, based on WHO	EF
Mexico	Substantial		14 years[e]		Local	EF, PP2
Morocco	Substantial	No		20 years	None	...
Nigeria	Substantial		20 years		WHO, United Kingdom, United States	EF for traditional medicines
Pakistan	Substantial		16 years		Yes	PP2
Philippines	Officially no, in practice, yes		17 years[f]		WHO	EF for traditional medicines
Republic of Korea	Substantial		15 years[g]		Voluntary	PP2
Sri Lanka	No direct product controls		15 years[e]		None	...
Sudan	Substantial		20 years		None	...

Country					
Syrian Arab Rep.	Substantial		15 years	None	...
Taiwan Province	No		18 years	...	...
Thailand	For essential drugs	No	15 years	WHO	...
Tunisia	Substantial	No		WHO	...
United Rep. of Tanzania	Substantial	Yes	20 years	WHO	EF for traditional medicines
Uruguay	Limited	No	15 years[e]	Local	...
Venezuela	Limited	No	5–10 years	WHO	...

Notes

[a] Unless otherwise indicated, patent protection refers to period from date of application.

[b] Excessive profits can lead to price modifications.

[c] Inadequate patent protection has affected innovative research.

[d] If profit margins are above or below the allowable margin of approximately 17 per cent, prices are modified.

[e] From date of grant.

[f] From date of issue of invention of patent.

[g] From date of publication of application.

PIC   Pharmaceutical inspection convention of European Free Trade Area.

FDA   Food and Drug Administration.

EF   External, mainly public funds, spent on health research.

PP1   Positive policy support to industry, measurable in financial terms.

PP2   Positive policy support to industry R&D, declared as intention/primary objective.

NP   Negative policy discouraging 'me too' drugs (therapeutic advance to existing therapy is condition for product registration) or encouraging the use of generic names and products.

Source: UNIDO, based on IMS and national sources.

well. The guidelines' purpose is to ensure that drugs are safe, efficient and stable. In order to do this, all phases of the manufacturing process must be monitored and controlled. This means that therapeutic ingredients (medicinal chemicals) and other inputs have to be of a consistently high quality, production facilities must be constructed and operated according to predetermined standards, staff should have specialized training and an elaborate system of records on operating procedures must be maintained.[5]

The guidelines are clear, although there is still wide scope for variation. WHO (1987) has developed an international set of recommendations for GMP which are referred to in Table 6.1. The WHO version is accepted by almost all industrialized countries, although many impose more stringent requirements (see Box 6.2). Only a minority of developing countries have adopted the WHO standards; most maintain local standards which are less strict or impose no requirements on their firms. One reason for the lower standards is that health authorities lack the resources necessary for regular inspection of production and distribution facilities. Another, more important, reason is the costs involved. Buildings, machinery and equipment, and air-conditioning and ventilation systems must all comply with

---

Box 6.2   Good manufacturing practices in industrialized countries

Large pharmaceutical companies in industrialized countries operate according to standards that are stricter than those recommended by WHO. In many cases they are required to do so by law. Box table 6.2 compares some of the main features of the WHO Certification Scheme with that of the United States as applied by the FDA.

Leading firms in industrialized countries go beyond the legal standards by incorporating various elements of 'total quality control'. Their motive in doing so is mainly to raise productivity and improve their competitive abilities. One example is 'just-in-time' (JIT) methods of inventory control, a technique which is widely practised in the automobile industry. JIT is used in the ordering and receiving of materials to ensure that they are available just at the time when required. With JIT, large inventories of materials become unnecessary. If the practice becomes widespread in the pharmaceutical industry, suppliers will be required to guarantee the quality of the materials received and must therefore employ compatible methods of quality control.

Box table 6.2   *A comparison of GMP standards of WHO and the FDA*

WHO	FDA
Status of GMP standards	
General guides, recommendation	Minimum legal requirements are specified; a drug is deemed to be adulterated unless it meets these requirements
Personnel qualifications	
Personnel should possess scientific qualifications, education and practical experience	Law requires that personnel be qualified by training and experience to perform the assigned tasks
Starting materials	
Starting materials and records should be kept of the supplier and the origin of materials if possible	Medicinal chemicals can be purchased only from known, reputable suppliers; the pre-approval necessitates an inspection of suppliers' facilities in many cases
Production and process control	
Documents relating to manufacturing procedures should be prepared for each drug; they should contain several types of information, including name and dosage form, evidence of stability and quality, theoretical yields and so on	Written standard operating procedures for each production process and control procedure are required by law; any deviation must be investigated; further processing is subject to approval of quality of semi-finished products after each significant manufacturing step; microbiological contamination is controlled
Quality controls	
The quality control unit should control all starting materials, monitor the quality aspects of manufacturing and control the quality and stability of drugs	The quality control unit has total responsibility for ensuring that adequate systems and procedures exist and are followed to assure product quality

GMP regulations. Supplies, too, cost more when purchased from approved vendors than from international trading houses. A third deterrent is that existing systems of price control rarely take into account the cost effects of GMP.[6]

National differences in manufacturing standards have important implications for trade patterns. Without mutual recognition of national guidelines, each aspiring exporter must first pass an inspection by the authorities of the importing country. Furthermore inspections must be carried out for each particular drug or medicinal chemical before it can be exported. This is the main reason for developing countries exporting a comparatively small portion of their production; most plants producing finished products cannot satisfy the inspection standards of the potential importer. Even when standards can be met, repeated inspections by each importing country are tedious and costly. As a result the bulk of exports from developing countries are commodity medicinal chemicals which are sold through international trading houses to other developing countries that do not insist on inspections.

The international acceptance of standards in industrialized countries has not progressed much further than in developing countries. Current and previous members of the European Free Trade Association (EFTA) have agreed on their own pharmaceutical inspection convention (PIC). If firms in one participating country satisfy domestic inspection procedures, they may export to other PIC signatories without submitting to another inspection by the importing country. Germany, Japan, Sweden and Switzerland also grant reciprocal recognition of each others' practices and a similar agreement exists between Sweden and the United States. Without an international agreement, multiple inspections are necessary before a firm can export certain drugs or medicinal chemicals.

The last column in Table 6.1 gives some idea of the financial and policy support offered to the industry. Most industrialized countries provide various types of direct or indirect subsidy, tax write-offs, and support for research which can be measured in financial terms. The industry also receives external funding – mainly public funds – to be spent on health research. Developing countries can rarely afford such largesse. Many governments are content to issue policy declarations in support of the industry but the tangible benefits of such statements are few. Those that do offer financial assistance generally designate it for support of research in traditional medicines. The alternative is to introduce policies to encourage the use of generic names and products or to require that each drug be a therapeutic advance as a condition for product registration.

The types of policies employed in industrialized countries are of particular interest, not only because they apply to the world's largest drug markets but also because they serve as models for many developing countries. The regulatory system of the United States is one of the more unusual among the industrialized countries. American pharmaceutical companies operate in an essentially free market and the government pays only a small share of the consumer's drug costs.[7] This is one reason for drug prices increasing more than twice as fast as the consumer price index during the first half of the 1980s.

The absence of price controls in the United States has spawned a number of indirect methods to limit drug charges. Cost-conscious health plans have sprung up to satisfy the growing demand for cheaper medicines. The most common of these is the Health Maintenance Organization (HMO) which provides a list of medical services and drugs that are paid for by a flat monthly fee. A policy of generic substitution (that is, the prescription of cheap copies of brand-name drugs whose patents have expired) is also pursued aggressively. Recent legislative reforms have reduced the amount of data which a generic producer must submit when applying for authorization of an off-patent product.[8]

The pharmaceutical industry in the EC is subject to a great deal of government intervention but the methods vary from one country to another. The lack of policy coordination is suggested by the fact that wholesale drug prices differ by as much as 300 per cent between the cheapest and most expensive markets. Many EC governments set a price before a new drug is launched and a few cling to the same limit throughout the drug's commercial life. In Spain and the United Kingdom officials take a different approach: they attempt to influence prices indirectly by controlling the firms' rates of return (see Box 6.3). Several countries supplement their system of price controls with a restrictive list of reimbursable drugs and then vary the proportion of drug costs that a patient can reclaim. Price freezes – either voluntary or compulsory – have been employed in a number of instances.

Political pressure for cheaper drugs is almost certain to intensify as the EC moves closer to its goal of a single market. Many Europeans

---

Box 6.3   Indirect methods of price control in the United Kingdom and Spain

The United Kingdom's Price Regulation Scheme (PRS) for pharmaceuticals specifies a permitted rate of return on capital. Drug firms are free to set their own price but cannot exceed a predetermined profit ceiling. The government began to reduce the industry's projected rate of return in 1983. Two years later the rate had fallen from 25 to 17 per cent. Officials subsequently restricted the industry's ability to charge promotional expenses against profits and limited the types of drugs for which the National Health Service (NHS) would pay.

Spanish officials adopted a similar procedure in 1991. Companies submitting a new product for pricing will have their return on capital limited to 12–18 per cent. Company assets, revenues, financial costs and all 'wholly justifiable' research costs are taken into account in the calculation.

---

want to see their national systems rationalized and coordinated. There is little reason for drug companies to apply separately to each EC member when seeking approval for a new product: the practice merely increases licensing costs which are already high. Some hope for a more centralized approach to approvals which would include an agency for reviewing the safety of medicines (see Box 6.4). The idea has drawn much criticism, however. These differences in national regulations and pricing schemes represent a barrier to foreign competitors and are a major reason for the European industry's fragmented structure. For example, individual governments offer companies better prices for their drugs if they invest locally in plants or research centres.[9] As a result the European map features far more pharmaceutical plants than would appear necessary.

European countries have also been slow to introduce policies to encourage the growth of generic markets. The overall share of generics in the EC's market is low – less than 10 per cent of total drug sales in recent years. Officials in the Federal Republic of Germany have been more open to this option than most. The country's Federal Health Office revamped its system for drug approvals in order to cater for generics and promote their use. The early results were striking: generics accounted for 15 per cent of total sales in 1987, compared with only 5 per cent in 1984. Later the pace of growth slowed when producers of original drugs cut their prices in response to the introduction of fixed-price supports for reimbursable drugs. Generics ac-

---

Box 6.4   An EC agency to regulate drug safety

The European Commission first tried to unify the drug licensing systems of the EC in 1985. That procedure, which allowed national authorities simply to recognize licences granted in other member states, did not work well. Not one of the 150 applications made under the system was automatically recognized by all other countries. The new proposal is for a European Medicines Control Agency (EMCA) to be set up after 1992. The EMCA would eventually have a staff of about 150 to administer requests from drug companies to sell new products in different countries. The current version allows for approvals in either of two ways. A company may submit an application to the EMCA. If the agency approves national authorities have 30 days to disagree and conflicts will be settled at the Community level. Alternatively a company may first submit its application to a national authority. If it is approved, all other national authorities must also approve and disagreements are again resolved at the Community level. The Commission hopes (perhaps over-optimistically) that by the second half of the 1990s most licensing will be handled by the EMCA and that national authorities will deal only with companies that do not market their drugs internationally.

---

counted for about 17 per cent of drug sales in the German market in 1990 and their share is expected to double by 1993.

The prices of drugs in Japan set this country apart from others and are mainly due to the unique methods of distribution which exist there. Doctors, rather than pharmacists, dispense prescription drugs. The country's national-health insurance programme sets a price for each medicine which the government pays to the doctor each time the drug is prescribed. Doctors, however, purchase their drugs at much lower prices from wholesale distributors and therefore can make a profit. In 1988, around 60 per cent of the earnings of general practitioners came from dispensing drugs. Japanese doctors have an obvious incentive to prescribe more expensive medicines than would otherwise be the case. Thus it is no surprise that the country leads the world in the amount of drugs prescribed for patients.

Policy makers in Japan have gradually begun to impose much tighter controls, reducing drug prices by 10 to 15 per cent every two years. They also require that some groups of patients pay a larger share of the costs of treatment. This more aggressive approach is an attempt to control the growth of costs in public health care but it stems from other factors as well. Japan's high drug prices were originally part of a strategy to support the pharmaceutical industry during its early stages. Gradually foreign producers have entered the Japanese market and domestic firms have begun to export. Because of these developments the practice of subsidizing the industry through higher prices is no longer so appealing (see Chapter 7).

Clearly the types of intervention practised in Japan, the United States and Western Europe are unique in various ways but the contrasts are even sharper when the policies of industrialized and developing countries are compared. These differences can be explained in terms of priorities of the government officials and in the degree of regulatory stringency they impose. For example, once attention turns to issues other than price controls, the goals of the two groups are markedly different. Those in industrialized countries are especially concerned about safety, efficacy and the provision of correct promotional information. Their counterparts in developing countries are more interested in promoting indigenous ownership, encouraging technology transfer and curbing the worst abuses of brand names and promotional claims.

The regulatory systems of developing countries are also less rigorous than those in industrialized countries. The reasons are fairly obvious. Some developing countries do not have the capacity to assess data on efficacy or safety. Many lack the ability to control the distribution of drugs or to enforce an adequate system of manufacturing guidelines through recurrent inspections of products and factories. Additional gaps in the regulatory systems can result from the fact that economic goals must sometimes take precedence over health objectives.

The importance which the developing countries place on price controls results from the fact that medicines are a 'basic need'. Drugs account for as much as half of all health care expenditures in these countries (compared with 8–10 per cent in industrialized countries). Any increase in prices is extremely costly in terms of the health benefits forgone. Policy makers therefore apply tight price controls, sometimes imposing price freezes for extended periods of time. When these freezes were lifted subsequently, drug prices soared. The effects are much the same in countries where pharmaceutical manufacturers have traditionally been permitted to import their raw materials at preferential rates of exchange. Later, when balance of payments crises forced governments to revamp their exchange-rate policies, the costs of production and prices of drugs rose substantially.

Developing countries frequently supplement their efforts at price control by attempting to economize on the purchase of drugs. They do this in several ways: by purchasing generic drugs, using centralized methods rather than going through conventional channels of distribution, by imposing especially tight regulations on the operations of foreign-owned subsidiaries or by refusing to recognize the patents on various products. These sorts of tactics pose few problems for local firms which conduct little or no research, though they are the source of much criticism from multinationals and governments of industrialized countries.

There are calls to reform regulatory systems in a number of developing countries, but these come at a delicate time. Many governments are simultaneously under pressure to expand their systems of public health care and the costs of such a move would be inflated if price controls were relaxed. Various groups, such as pharmacists and doctors, also occupy extremely strong positions in the distribution system and oppose the move. Such conditions occur throughout the world but the low degree of power and the limited ability of government officials to combat excessive influence is astonishing in some developing countries.

Recognition of the fact that regulatory systems in developing countries are often inadequate has led several international groups and organizations to assume a prominent role in the policy debate. The most important of these is the World Health Organization (WHO) which advises on drug strategies, drug policies and ways to set up or strengthen regulatory systems. WHO's efforts include its Action Programme on Essential Drugs, the Extended Programme on Immunization (EPI), the Tropical Diseases Research Programme and the Diarrhoeal Diseases Control Programme. In carrying out these and other programmes, WHO has the cooperation of United Nations Children's Fund (UNICEF), World Intellectual Property Organization (WIPO), the United Nations Conference on Trade and Development (UNCTAD), the United Nations Industrial Development Organization (UNIDO), the International Organization of Consumer Unions (IOCU), Health Action International (HAI) and other bodies.

In conclusion policy officials in developing countries face particularly harsh trade-offs between their desire to create an efficient set of producers and their need to ensure that the costs of medicines are not beyond the reach of the population. The following section examines some of the more specific issues which have figured most prominently in the policy debate in industrialized countries, while the subsequent one deals with a similar set of concerns in developing countries.

POLICY ISSUES IN INDUSTRIALIZED COUNTRIES

Policy makers in industrialized countries would generally prefer to leave many aspects of the industry's operation to market forces. Their desire, however, is tempered by clear evidence of oligopolistic behaviour among the larger drug producers. The view which prevailed in the 1960s and 1970s was that competition was 'good' while monopoly (and oligopoly) was 'bad'. The sorts of recommendations which followed from this puristical position focused on ways to stimulate competition, facilitate entry and enable the smaller manufacturers to obtain a greater share of the market. Policy makers now tend to adopt a more pragmatic approach. Rather than seeking to alter the market structure, they give more attention to issues of profitability, rates of return and basic operating conditions in the industry.[10]

Industry Profitability

At the heart of the debate is the question of whether profitability is excessive and, if so, the extent to which this can be attributed to the considerable degree of market power. The relationship between profitability and market power is not a straightforward one, however. Much of the industry's uniqueness stems from the high degree of interdependence which exists between research and distribution. The new drugs which research yields will become widely and rapidly available only if coupled with an elaborate network to provide information to prescribers (and to persuade them of the new products' superior therapeutic value).

In turn large-scale systems of distribution cannot be sustained unless they are coupled with massive research programmes. Distribution depends on the peculiar distinction between the 'choice-maker' (the medical practitioner) and the buyer (the patient or health scheme) and combines an inextricable mixture of information and persuasion. Some of the consequences are questionable. Examples are the creation of an 'image' for a new drug which allows it to command a higher price than rivals and the encouragement of brand-name loyalty so that the firm earns some profits from other, less novel drugs. However powerful distribution systems can also serve valuable functions. They provide practitioners with vital information and ensure that new products become available to large numbers of buyers as soon as local regulations are satisfied.

There is now a large body of literature focusing on profitability and related issues such as rates of return on research and marketing. Early studies confirmed that substantial differences exist between a firm's prices and its direct production costs. The original measures were faulted because they excluded outlays on research or marketing; the integral nature of these functions led analysts to adjust their methods to account for this fact. Even with these modifications the industry's rates of profit generally proved to be higher than the average for manufacturing.[11]

Spokesmen for the industry have responded to the accusation of excess profitability in several ways. First, they argue that the high prices charged for the limited number of successful products are necessary to compensate for the losses incurred on the many unsuccessful ones. Second, they maintain that, without substantial profits on marketable drugs, firms would be unable to fund research and marketing.[12] Finally, attention is drawn to the large number of products available in most therapeutic markets and the high rates of product introduction and obsolescence which occur regardless of rates of return.

What are the rates of profit in recent years? Table 6.2 provides some idea, showing the after-tax profit margins for 77 of the world's larger pharmaceutical companies. The estimates for individual companies vary widely, although the average for the sample (12.6 per cent) is probably higher than the corresponding profit margin for all manufacturing activities in industrialized countries.

Only a few countries report a sufficient number of observations to permit any meaningful comparisons. Profits in the United Kingdom and the United States are high and have remained so for several years. The relatively low profit margin of Japanese companies is also typical of the longer-term pattern but is somewhat surprising, particularly since that country's drug prices are rather high (albeit declining as regulations are tightened). Differences in profit margins may well be due to variations in product mix and types of specialization rather than differences in pricing regulations. Such a hypothesis, however, is

Table 6.2   *Average ratio of profit margin to total sales for major pharmaceutical companies in selected industrialized countries, 1984, 1987 and 1988*

Country	Number of companies	Average ratio of profit margin to total sales[a] (per cent)			Three-year average
		1984	1987	1988	
Belgium	1	8.4	7.4	8.1	8.0
Denmark	1	...	13.9	11.1	12.5[b]
Finland	1	...	11.5	13.8	12.7
France	4	12.9	12.0	10.5	11.8
Germany, Fed. Rep. of	1	2.9	2.5	1.7	2.4
Hungary	1	29.9	29.0	29.6	29.5
Italy	3	9.1	10.5	11.6	10.4
Japan	27	3.5	6.0	6.5	5.3
Netherlands	3	10.6	7.6	7.2	8.5
Portugal	1	...	3.2	2.2	2.7[b]
South Africa	1	...	12.9	14.6	13.8[b]
Spain	1	...	14.5	17.6	16.1[b]
Sweden	2	9.3	8.3	9.5	9.0
United Kingdom	4	21.7	25.1	27.2	24.7
United States	26	25.6	20.8	18.7	21.7
All the above[c]	77	13.4	12.3	12.7	12.6

Notes

[a] Profit margins are based on net after-tax income.

[b] Two-year average.

[c] Excluding the countries for which data on 1984 are missing.

Source: *Scrip Pharmaceutical Company League Tables* (various issues).

not verifiable; the number of observations for most countries is too few to draw any conclusions about the relationship between pricing policies and rates of return.

Questions relating to market structure and 'acceptable' rates of profit are unlikely to be resolved. Product competition certainly occurs (see Chapter 5) and few firms, if any, enjoy a true monopoly position. Yet there is ample evidence to suggest that large firms retain a substantial degree of market power. Average returns for NMEs may have declined over the long term but a number of new products generate impressive profits and the industry's performance continues to exceed the average for all manufacturing. Moreover entry is negligible and the leadership of major firms has remained unchallenged over long periods of time.

All these circumstances imply that market power is sometimes great. The inability to resolve this issue is one of the reasons for the scope of the policy debate having narrowed, focusing on the industry's uniquely interrelated systems of research and distribution. The remainder of this section discusses some of the relevant policy implications in these two key parts of the industry.

Rates of Return on R&D

The effectiveness of the industry's research programme has drawn much criticism. Several of the harsher critics argue that the huge amounts spent on research have yielded comparatively few social benefits. They maintain that most new discoveries actually come from sources outside the industry, while the companies' own laboratories are primarily concerned with 'molecule manipulations'. Known as 'me too' drugs, these are therapeutically similar to medicines which are already on the market although they are still regarded as new products. Such unnecessary diversity increases risk and cost, since 'me toos' are generally marketed as superior versions to their predecessors and therefore are more expensive.[13] Spokesmen for the industry acknowledge that the first molecular variation often does not originate in its own laboratories. However they do not concede that later versions will necessarily fall into the 'me too' category. The first version frequently proves to be inferior to later variations, many of which emanate from industry laboratories.

Other critics concede that the spending on research and product development has brought certain benefits, although they question the usefulness of research in certain parts of the industry. This interpretation is often transposed into a debate on whether large or small firms are the better vehicles for technical advance. Regulators have argued that the research of small firms represents the more significant contribution to product development. By this they imply that, even if large firms were forced to cut back on research spending (as a result of more stringent regulations on pricing and testing, for example), the industry's research inputs and outputs would not be diminished. Studies based on data from the late 1960s provided support for this view, suggesting that a firm's research effort (measured by the share of R&D in total expenditures) tended to diminish as the company grew larger (Mansfield, 1968, pp. 38–40; Grabowski, 1968, p. 304).

Later analysis cast doubt on this interpretation, however. According to one line of investigation, the decline in a firm's research intensity can be deterred by high levels of internal cash flow (Grabowski and Vernon, 1981, pp. 14–15). Other results underline the trade-off between the need for stringent regulations and the desire for rapid research progress. Increased regulation, for example, seems to have a particularly adverse effect on the smaller producers. These firms at present account for a lesser fraction of industry-wide R&D than in the early 1960s (Thomas, 1987).

The question of the relationship between firm size and research efficacy may soon be decided by the industry itself. Even the strongest proponents of large-scale operations concede that there may be little to gain by building up big research operations. A more promising strategy is to rely on small-scale research and large-scale development (for example, the performance of all the routine tests needed to meet regulatory requirements). The important point to note is that much of the problem results from conditions within the industry and cannot necessarily be attributed to a stronger regulatory environment.

Another part of the debate focuses, not on the means of productive research, but on the rates of return it generates and whether these are sufficient to cover the costs of R&D. Studies based on data for the 1950s and 1960s found that rates of return for R&D were substantial – between 30 and 45 per cent (Pakes and Schankerman, 1984, pp. 73–4). Subsequent investigations revealed far lower rates of return. The general consensus which emerged was that returns on research spending probably reached a peak in the early 1960s and fell steadily over the next two decades (see, for example, Baily, 1972; Statman, 1983).

These findings were regarded as evidence that more stringent regulations had a negative impact on product development. However the latest information suggests that rates of return on research have begun to rise again and are now roughly equal to the industry's cost of capital. Grabowski and Vernon (1990) attribute the improved economic performance to three factors. First, the industry has been able to boost its research productivity. R&D is increasingly characterized by a 'discovery by design' approach, rather than the type of random screening which was common during most of the post-war period. Second, researchers are focusing more on chronic rather than acute health care problems and this raises the rate of return. Third, there has been a change in pricing practice. Drug firms were raising their prices more rapidly than inflation during the 1980s and this allowed them to cover investments in R&D. A growing number of companies have apparently been able to adjust their research strategies to the more stringent regulatory environment which exists today. This impression is supported by the steady growth in real expenditures on R&D, which increased by about 10 per cent per annum throughout most of the 1980s, or roughly double the growth rate of the previous decade.[14]

Despite a large body of research, the trade-off between a more carefully regulated industry and a slower pace of product development remains unclear. Nevertheless there are grounds to conclude that controls aimed at maintaining certain standards of efficacy and safety need not jeopardize research progress provided that firms are sufficiently flexible to accommodate these policies.

Advertising and Promotion

The criticisms voiced with regard to distribution activities reflect many of the same broad concerns as those expressed in the case of research. The social gains to be derived from the huge amounts spent on these activities are questioned. Such outlays, which inevitably lead to higher prices and increased market power, are suspect on these grounds as well. The industry's defence is a fairly standard one. Spokesmen insist that outlays on advertising and promotion supply essential information to physicians and other choice-makers. Given the large number of drugs introduced every year, no other source of new information is readily available.

The way in which advertising and promotion might alter competitive conditions is based largely on impressionistic lines of reasoning rather than empirical evidence. The essence of the critics' argument is that major pharmaceutical companies provide a great deal more information than doctors can absorb. At some point the information flow becomes so great that it inhibits the entry of new producers: a new entrant would have to make prohibitively large outlays on advertising in order to displace the recognition accorded to established products. Small firms that cannot afford this scale of spending are denied access to the physicians who make purchasing decisions. A number of studies have cast doubt on specific parts of this argument, though their results depend critically on the way competition is defined and measured.[15]

A more productive line of investigation is indicated by the work of Bond and Lean (1977). They argue that physicians tend to respond most favourably to the promotion of brands which are the first to offer a therapeutic advantage. Such a finding implies that the relation between promotion and sales is not only brand-specific but depends crucially on the timing of the innovation. In terms of market power the essential question is whether high levels of advertising spending result in (i) increased sales or (ii) higher prices than could otherwise prevail without depressing demand.

There is little doubt that pharmaceutical advertising provides considerable information to choice-makers on the characteristics of new and established drugs. But controversy – even within the confines of this statement – persists. The prominent role played by advertising may largely reflect competitive pressures, although the huge amounts spent on promotional campaigns suggest that other objectives are also served. In the United States, for example, promotional expenditures are roughly equal to total sales in the first year of a product's launch. They decline to 50 per cent of sales in the second year and 25 per cent in the third (Grabowski and Vernon, 1990). The sheer volume of advertising, along with the role played by the detailmen who carry the message to physicians, seems to call for a rather comprehensive system of regulation and control.

POLICIES IN DEVELOPING COUNTRIES

The fact that pharmaceutical policies in developing countries have their own distinct flavour has already been noted. The limited buying power of consumers, the fragmented nature of public health care and a shortage of research skills all influence the choice of policies. Pharmaceuticals also have a 'foreign' dimension which is probably greater than that of most industries. The issue of foreign versus domestic ownership is one which figures prominently in the developing countries and is discussed here, along with other aspects of pricing, marketing and patent agreements.

Controls on Foreign Ownership

Foreign-owned firms are typically in a minority in developing countries but they are much larger than their domestic rivals and frequently account for well over half the home market (see Table 3.7). Critics of the multinationals believe that such a substantial foreign presence has undesirable consequences. Their view is sometimes inspired by 'nationalistic' rather than economic considerations,[16] but a locally-owned industry in close touch with the needs of the domestic market can offer certain economic advantages. Local producers, for example, may be more willing to compete in generic markets than foreign subsidiaries and, presumably, would be more adept at identifying those product markets where demand is growing. They should also be more willing to distribute their drugs outside the major metropolitan areas, supplying generic drugs to users in poor or rural areas.

Whatever the reasons, policy makers in developing countries have attempted to alter the existing pattern of ownership in two ways. Some have insisted that the degree of foreign ownership in equity shares be scaled back to a predetermined maximum (usually 50 per cent or less). That has been the approach in India, where both the number of foreign firms and their share of the domestic market is less than a quarter. Other countries have followed a less direct path. They encourage the development of an indigenous industry by implementing various types of joint ventures and licensing agreements, by controlling methods of drug procurement, by encouraging the transfer of technologies, or by placing restrictions on promotional campaigns to limit the influence of the brand-name system (see Box 6.5).

Attempts to encourage indigenous production through the international transfer of technology have led to a particularly acrimonious debate involving the multinationals as well as government officials in industrialized and developing countries. New products are disseminated fairly rapidly among the developing countries but the innovators rarely have any incentive to ensure that the same happens with product or process technologies. These transfers may nevertheless be essential,

---

Box 6.5   Policy support for local firms in developing countries and areas

The alternatives which are available to governments in developing countries are many, depending on conditions in the home market. Drug markets in several parts of Asia are growing rapidly and attitudes towards foreign investment are comparatively liberal. Governments in Malaysia, the Republic of Korea, Singapore, Taiwan Province and Thailand therefore rely heavily on joint ventures and licensing agreements in an effort to promote their domestic producers. The Republic of Korea reported more than 30 joint venture agreements with multinationals at the beginning of 1989.

In the Philippines, buyers depend heavily on brand names and this works to the advantage of multinationals. One result is that the domestic industry imports up to 95 per cent of its raw materials. In order to build up an indigenous industry, the country's national drug policy aims at controlling the influence of brand names and encouraging the local production of basic chemical raw materials.

The pharmaceutical industry in most African countries is still at an early stage of development and the types of policy alternatives differ accordingly. Members of the African Preferential Trade Area (PTA) are attempting to replace some of their imports from multinationals with increased purchases from PTA firms.

---

particularly when other regulations are imposed to curb the domestic presence of multinationals. Acquisition of product and process technologies usually requires that the government circumvents or ignores the intellectual property rights granted in industrialized countries. Those developing countries which have been major participants in the disputes about technology transfer already have a relatively sophisticated pharmaceutical industry. Many of their supporters have argued that patent-protected drugs are simply too expensive in comparison to locally-made copies which sell for as little as a quarter or a third of the price of a brand-name product.

Multinationals have responded with several arguments. They argue that many of the local firms which emerge with the help of domestic policy are often inefficient and will remain so because international competition is excluded. Another line of defence is to suggest that the exclusion of foreign firms jeopardizes the quality of products since the technologies used by domestic producers will not be the most advanced or efficient. Finally industry spokesmen insist that the failure to recognize patents provides drug counterfeiters with a legal base from which to supply their products.

Distribution Practices

Dissatisfaction with some of the consequences of marketing and promotional campaigns has been voiced in almost all countries. Many of the critics' charges are similar to those heard in industrialized countries, but the consequences are sometimes more acute in view of the relatively lenient regulatory systems and the developing countries' shortage of scientific and medical skills.

The arguments against these over-zealous practices are well known and need only be briefly summarized here. First, today's distribution systems are costly and can result in drug prices that are unnecessarily high. Second, consumers in developing countries are exceptionally brand-conscious and large-scale advertising campaigns will create privileged market positions for multinationals. In such circumstances the introduction of cheap generics becomes difficult and this may deter the entry of less innovative, local firms. Third, the limited number of pharmacists and medical practitioners increases the danger that information obtained through promotion and advertising will be misleading or interpreted incorrectly. Finally there is a possibility that firms will sell drugs which are dangerous or even banned in industrialized countries. The multinationals themselves have developed a code of pharmaceutical marketing practices to guard against this eventuality, but the lapses which have occurred have led some governments to question the value of the entire promotion system.

In summary the key issues in the debate about marketing practices revolve around the brand-name system – the way it affects prices, the possibility that it inhibits competition from local enterprises and the fear that it could be open to abuse in extreme cases. Policy makers therefore have little choice but to attack this cornerstone of the industry, even though their ultimate objectives are much broader in scope. In order to reduce prices they seek to encourage the growth of generic markets, but that goal requires that the prominence of branded drugs be scaled back. Increased competition can also result, since local firms selling under generic names will find that market entry becomes easier. Multinationals, it is hoped, will respond ·by launching their own generics and this, too, would lead to greater competition.

Policy makers must nevertheless be cautious in seeking to replace the brand-name system with a host of generic drugs. The development of generic markets will cut the cost of medicines, but, if carried too far, it can have negative effects. Quality controls must not be lax. Nor must the market presence of the multinational be drastically reduced: consumers might then be denied ready access to the latest products and innovations which are rapidly distributed elsewhere in the world.

Drug Pricing

Government officials have two choices when devising policies to deal with pharmaceutical prices: they can either impose direct controls or allow firms a measure of freedom in setting prices and then reimburse a portion of the consumer's costs.

Table 6.3 examines the approach in a number of countries. Controls may be employed at the wholesale level, at the retail level, or both. Price changes may also be subject to government approval, although in many cases the controls are not general, being more stringent for essential drugs, drugs supplied by the public sector or drugs sold by foreign firms. Price freezes are sometimes employed in conjunction with controls, while a number of other variations can also be noted. From this information it is clear that most governments prefer to intervene in favour of the consumer by controlling prices rather than offering subsidies. The low per capita incomes in most developing countries, coupled with the relatively high prices of medicines, make the former option more attractive.

The developing countries' motives for such intervention are similar to those of industrialized countries and need not be repeated here. However the fact that multinationals dominate these markets so completely gives rise to unique problems. These include the issue of transfer pricing and the inherent conflict between the patent system and the goal of obtaining modern drugs at minimum prices. Transfer prices refer to intra-firm sales between affiliates of multinationals located in different countries. To the firm they may be regarded purely as an accounting mechanism. However, when the intra-firm prices of pharmaceutical intermediates exceed the price charged to an unrelated concern, the profits of the selling subsidiary are increased. Similarly the under-pricing of intermediates transfers additional funds from the seller to the buyer even though both may be part of the same multinational corporation. Such practices are common to many industries but the gap between transfer and arm's length prices tends to be larger in the case of pharmaceuticals than in others (Gereffi, 1983, p. 195; see also Pradhan, 1983, pp. 127–32). Transfer pricing is practised everywhere but it can pose a serious problem in a developing country where the multinationals' domination of the domestic market is almost complete (see Box 6.6).

Critics of the practice argue that its purpose is to take advantage of differential rates of taxation which exist in the host countries of the buyer and seller. The industry's response is that a large portion of its 'global costs' are incurred at headquarters, where R&D, drug testing and production facilities are usually located; the price structure of headquarters' sales therefore differs from that of subsidiaries and transfer pricing represents a means of distributing these costs across the firm's network of affiliates, all of which benefit from the centralized activities.

This argument meets with a rather cold response from many government officials. Comparatively little research is conducted on diseases which are unique to the developing countries (research on tropical

---

Box 6.6   Transfer pricing in Pakistan

The issue of transfer pricing has been a recurrent one in Pakistan, where 32 multinational companies accounted for 70 per cent of the domestic drug market in 1989, with the remaining 30 per cent being shared among 150 local manufacturers. Because most multinational subsidiaries operate by processing ingredients imported from their parent companies, transfer prices that are fixed at high levels will have widespread consequences. Pakistani sources have estimated that this practice has pushed up the import bill by as much as 500 million rupees in 1989. Whatever the case, the estimates in the following table indicate that some of the prices charged for raw materials are much higher than world prices.

Box table 6.6   Price comparisons for pharmaceutical raw materials, 1986–7

Raw material	Pakistani price for imports ($/kg)	International price ($/kg)
Cimetidine	550	167
Doxycycline	1 750	550
Trimethoprim	315	35

Source: Scrip World Pharmaceutical News No. 1516 (1990) and Pakistan Pharmaceutical Manufacturers Association (1989).

Table 6.3   Price controls in developing countries

Country/area	Existence of price controls	Level at which price is controlled	Price changes subject to control
Algeria	Yes		
Argentina	16% of market is controlled	At manufacturing and retailing levels	
Bangladesh	For 45 essential drugs required for primary health care	Retail	Yes
Bolivia	Yes, since 1982		
Brazil	Partial; on permanent or continuous use drugs	Wholesale Retail	Price council sets increases.
Chile	No		
Colombia	Yes	Wholesale Retail	Yes
Costa Rica	Yes	Wholesale Retail	Yes
Ecuador	Yes	Wholesale Retail	Yes – a devaluation of more than 10% can bring price increases
Egypt	Yes		
Guatemala	Yes	Wholesale Retail	Yes
Hong Kong	No		
India	Yes	Retail	All changes require approval
Indonesia	Yes	Intended price to be submitted at registration	Yes

General or specific controls	Measures of control	Other features
	Monopoly over manufacture, import and distribution	Domestic prices are lower by a third than in France
Essential drugs (6% of market) are strictly controlled		Some prices freed in 1990
General, including finished goods, raw materials, packaging and intermediates	Retail price list and a price list for imported goods	Any new policy is likely to decrease price and encourage domestic manufacture
Local firms get better prices than foreign firms.		After a long period of controls prices have been freed for most drugs, but maximum prices are being published.
	Price freezes in 1985	Since 1986 some increases have been granted
Price increase of 45% for local and 15% for imported goods allowed		Cost analysis and prices of competing products are used to determine prices
All drugs	Periodic price freezes	Prices of raw materials were fixed in 1987; similar prices had to be charged for similar active ingredients
Local and imported drugs controlled		The pharmaceutical industry was nationalized in 1962–3
Price increases of 45% for local products and 15% for imported products allowed	Retail price list of available pharmaceuticals	In mid-1990 government began to tighten controls
Control of drugs used by Ministry of Health and the essential drugs list	Price freezes Extension of controls to more drugs	Pharmaceuticals researched in India are free of controls for five years
Differentials exceeding world prices by 10% are unacceptable unless explained	Official retail price list	Several price increases granted in 1986; Ministry claims Indonesian prices are higher than prices in other ASEAN countries

Country/area	Existence of price controls	Level at which price is controlled	Price changes subject to control
Jordan	Yes		
Kenya	No		
Libyan Arab Jamah.	Yes		Yes
Malaysia	No		
Mexico	Yes	Registration	Yes, lowest priority given to multinationals
Morocco	No		
Nepal	Only in specific cases		
Nigeria	Yes	Retail	More than a 5% increase disallowed
Pakistan	Yes	Registration	Yes
Peru	Yes	Registration	Yes, favourable to domestic firms
Philippines			
Republic of Korea	Yes	Wholesale Retail	Yes
Saudi Arabia	Yes	Registration	Price increases difficult to obtain
Singapore	No		
Syrian Arab Rep.	Yes		
Taiwan Province	No		

General or specific controls	Measures of control	Other features
		Factors determining prices are: prices in country of origin, in other Arab markets, prices of similar products, level of research and scientific standards of manufacture
		According to *Pharmacy World Journal* retail prices increased by 80% during 1979–87
Pharmaceuticals with a cheaper equivalent are unlikely to be granted an increase	Price freeze since 1972	
		Manufacturers permitted to set prices
More rigid for social sector	Price freeze since 1988	A 7% price increase was granted in January 1991 and industry is hopeful that all prices will soon be freed
		Industry pressure to raise prices
		Locally produced pharmaceuticals are 20–25% cheaper than imported goods System of voluntary controls in existence
		In practice controls are limited; a short supply of drugs has caused high price increases of up to 44% in 1980–5
General	Maximum prices Price freezes	Index-linked price rises have been granted for domestic products and prices of imported drugs will depend on the rupee–dollar rate
		Prices of imported drugs should not be higher than prices in other parts of the world
	Price monitoring. In 1985 prices of 300 drugs were lowered	Proposed price must be submitted at registration
Greater controls for drugs on National Health Insurance reimbursable list	Maximum prices Price freezes Price decreases	Price variations among new and existing products and prices of imported materials should not be too high
Prices of imported products are reviewed yearly		Similar products must have similar prices in the domestic market and not exceed wholesale prices in country of export
		A standard drugs list of 500 products for use in government hospitals and clinics uses price as one factor for selection of the drugs
		Prices in Jordan, Lebanon and in the country of origin are used to fix prices

Country/area	Existence of price controls	Level at which price is controlled	Price changes subject to control
Thailand	Partial		
Tunisia	Yes, strict control	Registration	Yes
Turkey	Yes	Registration Wholesale Retail	Yes, 10 day's notice and proof of rising costs to be given
Uruguay	Not for new products		Yes
Venezuela	Partial; for 40 basic drugs		Yes, only for the 40 essential pharmaceuticals
Yugoslavia	Yes	Wholesalers' purchase price	

Sources: UNIDO, based on Scrip and industrial sources.

diseases, for example). The result is that a portion of the price charged to consumers in developing countries presumably pays for research costs on diseases which are most commonly found in richer countries.

The second issue of unique significance results from the fact that modern, patented drugs are usually required to treat the full range of diseases which occur in developing countries. Generic drugs are the preferred source of pharmaceutical supplies, but when new and more effective medicines are needed government officials will usually look for suppliers (local and foreign) that do not observe patents. In that case special problems arise. Countries that practise this approach argue that consumers in industrialized countries should be expected to pay more for innovation than those in developing countries. They also insist that drugs intended for primary health care should be priced lower in developing countries than in industrialized ones. Some advocates allow for a distinction between public and private markets: drugs used in public health programmes should be set at acceptably low prices although those distributed through private channels can be sold at higher prices, which allows some rate of return for innovation.

In conclusion the policy objectives of industrialized and developing countries are rather distinct although, over the longer term, this

General or specific controls	Measures of control	Other features
Prices on essential drugs list		
	Profit margins of importers and manufacturers controlled	
		Price increases are linked to inflation and granted monthly
Public-sector institutions pay lower prices		Generic versions encouraged in order to lower prices; the 1985 average price of a drug was US$ 0.86, compared to the Latin American average of US$ 1.48
Manufacturers can set a price up to 30% of foreign price of product	Price freezes	In June 1990 an average 32% price increase allowed, but this is not a free market price

gap should narrow. The desire to control prices has always been the overriding goal of policy makers in developing countries. This objective now receives a higher priority in industrialized countries as well. As the costs of public health care rise, the search for new ways to limit prices increases without seriously jeopardizing research capabilities will widen. Officials in industrialized countries have an advantage in pursuing this goal: their consumers are more open to a policy of generic substitution and are better able to take advantage of tactics such as cooperative buying and other cost-saving measures.

So long as brand-name loyalty remains a prominent feature of most drug markets in developing countries a policy of generic substitution will face resistance. Such a move is strongly opposed by the multinationals which have invested heavily to promote the brand-name system. The prospects for compromise may improve in the future, however, as more developing countries introduce some elements of patent protection.[17] With this concession, multinationals might be more willing to accept some weakening of the brand-name system. Once the market for generic products begins to grow, the multinationals themselves could launch their own versions (with and without brands) just as they do now in industrialized countries.

Some of the policy implications of these developments are examined in the concluding chapter of this publication, but before considering these the industry's own responses to various forms of policy intervention and other market developments should be taken into account. The following chapter looks at a number of different corporate strategies which began to emerge in the late 1980s and early 1990s.

NOTES

1. Thalidomide, a popular sedative, was sold in 46 countries between the mid-1950s and 1961. Belatedly it was realized that the drug was responsible for birth defects in approximately 12 000 children.
2. For example, in the United States 'good laboratory practices' (GLP) which specify methods of documentation and evaluation of research results were not formalized until 1971.
3. Drug counterfeiters operate in both developing and industrialized countries. Annual revenues lost by American firms are estimated to be between $3 and 5 billion. World-wide the costs would be much higher.
4. Fundamental public health policies concerned with the regulation of drug quality – product licence for sales, registration of manufacturing premises with health authorities, classification of prescription and OTC drugs – are discussed elsewhere in this publication.
5. A master file must be maintained for each drug which describes standard operating procedures for each step in the process of production and control. Many of the documents and records generated during the manufacture of pharmaceuticals must be kept for review for at least five years following the date of manufacture.
6. The recent experience of Indonesia provides some idea of how GMP regulations may affect the industry. Small pharmaceutical firms in that country expect the rehabilitation cost of compliance to be around $270 000 (*Scrip, World Pharmaceutical News*, 25 July 1990). A series of mergers and acquisitions will have to occur as part of this process since many small firms cannot afford these outlays.
7. The government reportedly pays only about 1.7 per cent of identified drug costs. The actual share is higher but is not known precisely because the drug costs of hospitals are not itemized (*Scrip, World Pharmaceutical News*, no. 1462, 1989).
8. The move was part of a wider effort of legislative reform in the United States which entailed several compromises. The new laws helped generic drug manufacturers but also extended the period of patent protection for medicines.
9. Such practices are most common in France and Italy. Canada, too, has introduced a new law which grants companies patent protection for new compounds in return for increased research investment in the country.

10. The shift in emphasis can be attributed to two factors. First, many analysts have concluded that the tools for gauging monopoly power are either inadequate or inappropriate in the case of the pharmaceutical industry (see Comanor, 1986). Second, some regulators now concede that a substantial degree of market power is a necessary condition (although still subject to regulation) if the development and distribution of new drugs is to proceed at a satisfactory pace.

11. Temin (1979) has shown that adjusted rates of profit are more than two standard deviations greater than the mean for all manufacturing.

12. Recent studies of firms in the United States have shown that mean costs are covered for only the top 30 drugs when R&D is calculated on a fully allocated basis (Grabowski and Vernon, 1990).

13. For this reason some governments recommend that doctors prescribe from a limited range of the more cost-effective drugs. Patients who want more expensive versions must pay for them. This approach is followed in the Federal Republic of Germany, for example.

14. This evidence of a turnaround in research performance is based on the work of Grabowski and Vernon (1990) who refer only to the United States industry. Similar tactics were adopted by firms in other industrialized countries, however, and an improvement in rates of return to research is also likely in that case as well.

15. For a survey, see Comanor (1986).

16. The tendency to base industrial policies on nationalistic rather than economic considerations is, of course, not restricted to developing countries. Most industrialized countries single out certain industries as 'national champions' and adopt policies which are designed to foster the expansion of these industries. Industry examples include computers, electronic components, aerospace and advanced materials.

17. Brazil, Chile, Mexico and Thailand are examples. Brazil and Mexico have announced plans to introduce patent protection; in 1990, Chile passed a law recognizing patents for 15 years and Thailand has promised to introduce patent legislation in 1992.

7. Strategies of Firms and Industries

The pharmaceutical industry has long regarded itself as in some ways unique. There is some justification for this aloofness. The industry's reputation is partly based on the fiercely independent character of its firms and their secretive methods of operation. Even more important, pharmaceutical companies have proved to be exceptionally profitable and highly innovative. They have attained this record while avoiding much direct competition and relying very little on cooperative alliances. Finally drug companies have collectively managed to deter the entry of outsiders. Until recently there had not been a new entrant in the industry for decades.

These hallmarks of the industry are gradually being eroded. Drug producers are becoming involved in mergers and acquisitions, recruiting new managers, reducing size and restructuring. And they are competing much harder with rivals. In short the pharmaceutical industry is losing some of its uniqueness and becoming more like other parts of the manufacturing sector. The industry's transformation is mirrored in the strategic decisions of firms – the way they choose to compete and collaborate.

In one sense the types of strategic decisions required of any firm are fairly simple and straightforward. The most fundamental issues concern the products to be produced and the consumers or market segments to be served. Other decisions that follow from these deal with the choice of production processes and channels of distribution. The range of choices becomes more complicated when it is recognized that managers can vary the priorities they assign to any of these operations. Firms that wish to have an innovative and original line of products need not always be among the leaders in research. In fact there is an increasing tendency for R & D to be decoupled from other parts of the industry as companies buy into projects of various sizes and at various stages of completion. Methods of distribution will also differ, not just from company to company but from product to product. Each drug producer must decide whether to distribute its entire range of products, to allow outsiders to handle the distribution of certain products, or to turn all distribution responsibilities over to specialized firms.

The priorities which a company assigns to various countries and markets further enlarges the menu of strategic options. A joint venture may be the best method to obtain market access for OTCs, although a wholly-owned division could be needed for distribution of products through the medical profession. Markets can also be served by a joint venture that licences the research of several pharmaceutical firms and then distributes the drugs through channels owned by a specialist wholesaler.

The number of strategies that can be constructed from all these options is large, though in practice only a few are actually used. This chapter focuses on strategic decisions relating to research and distribution, the two most important parts of the pharmaceutical industry. However the two stages should not be seen as independent operations, with distinct activities having their own set of strategic options. In fact one of the growing concerns of many pharmaceutical companies is to ensure that they obtain a proper 'fit' between research and distribution. Managers seek to optimize the alignment between the type of R & D they carry out and the type of distribution system they use. Failure to achieve this alignment means that costs rise too quickly or that market segments will not be adequately served at any level of production.

RESEARCH-BASED STRATEGIES

The competitive standing of large, integrated producers depends mainly on their access to marketable products; those that do not plan their research activities wisely will contract or disappear. Meanwhile the growing costs of R & D make it more difficult to finance the big research programmes that were so common in the 1980s. As these costs rise, so too will the risks associated with R & D. The richest companies will still be able to minimize these risks by simultaneously funding a full range of research projects. Others will be forced to consider a narrower range of choices. This trend is already apparent among biotechnology companies, but mainstream pharmaceutical firms will soon have to face similar decisions.

A successful research programme is crucial for integrated producers but the way in which it fits into a firm's overall strategy is analogous to that in other, less research-intensive industries. The most fundamental issue is the familiar 'make-or-buy' decision. In the case of drug research, the circumstances surrounding this decision are particularly complex. The firm may decide to acquire a finished product (for example, a completed research project) or it may buy research which is still under way. The purchase price of completed projects will reflect the certainty that the research has 'paid off'. Various criteria for efficiency and safety will have been satisfied and the requirements of government approval will have been met. When research is purchased before these tests have been successfully completed, the price will reflect these uncertainties – the chemical compound, for example, may prove to be inefficient or fail to meet government requirements.

The make-or-buy decision has widespread repercussions, affecting opportunities for inter-firm collaboration, the types of marketing and distribution tactics to be employed, and the nature of the products to be sold. Various industry-wide scenarios can be constructed, depending mainly on how pharmaceutical companies choose to deal with this issue. One foresees a breakup of the industry into three distinct groups. The first group would be composed of producers with a pronounced research orientation. In the second would be firms that retain only a minor research emphasis. These companies, however, could still obtain new drugs through licence from research specialists and offer a full line of products to the market. The third major group would consist of producers supplying generic products.

Analysts subscribing to this scenario cite the recent experience of several large companies which have sought to retain a full range of research activities while continuing to market their own products. As governments in major markets have tried to persuade doctors to prescribe generic rather than branded items, some of these firms have experienced difficulties. Sales volumes declined and sales costs per outlet rose because more and more resources had to be expended to convince doctors of the merit of branded drugs.[1]

This sort of industry fragmentation should eventually create additional opportunities for specialization. One possibility is that more companies will begin to specialize in patent brokering – that is, finding markets for drugs which other firms have developed but do not consider worth pursuing.[2] Small pharmaceutical firms that lack the resources to conduct their own market search have traditionally been the major users of such services. In the future, large integrated firms will also seek help in finding markets where they can sell their patents. Stricter enforcement of patent laws can be expected and, as it develops, the need for patent-brokering services will grow.

A second scenario is advocated by analysts who expect some vertically integrated firms to reorganize, converting themselves into dual-line producers of generics and on-patent drugs. Methods of marketing and distribution could become major sources of competitive differentiation if drug makers move away from the traditional model of a highly integrated company. Such a development would also lead to much greater fluidity within and between research laboratories. Each project would then have to be judged not only on its technical promise but also in terms of the 'fit' it provides for the firm's overall competitive strategy.

No single scenario will dominate among the larger integrated firms. That would not be possible in any case; if a significant majority should adopt very similar strategies, competition would be too intense and a number would be doomed to failure.[3] The knowledge required for particular markets is nevertheless very specific and becoming more so. Meanwhile regulations are changing quickly and further complicating the distribution process. These trends will force firms to experiment with new types of tactics and organizational methods. Those which emphasize research and product development will find it advantageous to turn over some of their marketing responsibilities to others such as patent brokers.

Whatever the strategies adopted, this increasing balkanization of the industry will give rise to greater opportunities for collaboration in an industry which, until now, has been noted for its secrecy and independence. Drug producers everywhere will become more accustomed to using some portion of external inputs and/or services to supplement their own in-house capabilities. In principle firms may choose from a variety of options involving research, distribution or brokering. Several of these alternatives would be precluded, however, once a decision was made on the area of specialization or 'focus'.

MARKETING STRATEGIES

Decisions with regard to the firm's research orientation will have significant consequences for its marketing operations, in particular, the end-user channels (retail outlets or doctors) through which drugs are sold. Producers have come to recognize this interdependence, though the approach of many still tends to be guided by two long-held assumptions.

The first of these is that marketing activities benefit greatly from economies of scale. The importance attached to the role of scale economies is evidenced by the recent consolidation of companies (discussed below) and the steady growth of the sales forces in major firms. Estimates for the United States show that drug salesmen made nearly 30 million calls on office-based doctors in 1989. The figure represents an increase of almost 50 per cent over 1982 and this occurred without any rise in the number of American doctors.

The second key assumption is that integrated companies can expect profitability to fall if they venture into the production of generics. The basis for this assumption is that prices of generics are low, with gross margins of only 5–10 per cent, compared to 40–50 per cent for on-patent drugs. However there are several reasons to explain why this assumption may no longer be valid today. First, the per-treatment price of generics may be low in comparison with on-patent drugs, but so, too, are the costs. Second, generics require little in the way of additional research expenditures; by definition, these costs were incurred in creating the patented version. Third, manufacturing costs are less than half of total costs and, if judged on an incremental basis, a generic drug's marginal manufacturing cost will be low. Finally, the need for an elaborate marketing apparatus will gradually diminish as more and more generics are prescribed by doctors or are bought over the counter in preference to branded alternatives. Factors such as these explain why many drugs are still profitable long after their patents have expired.[4]

The industry's attraction to generics depends on other developments as well. One is the imminent expiry of patents on many popular drugs. In the United States, for example, roughly 80 of the 100 best-selling prescription drugs had come off patent by 1990, creating new opportunities for producers of generics.[5] Table 7.1 shows that the patents and patent extensions on several more major drugs will have

*Table 7.1   Major drugs losing market exclusivity in the United
States by the mid-1990s*

Maker	Drug	Use	Expiration	Estimated sales in 1990 (US$ m.)
American Cyanamid	Piperacil	antibiotic	1995	145
American Home Products	Nordette	oral contraceptive	1991	190
	Ovral	oral contraceptive	1991	140
	Triphasil	oral contraceptive	1993	100
Bristol-Myers Squibb	Corgard	heart	1993	125
	Capoten	heart	1995	580
Ciba-Geigy	Lopressor	heart	1993	215
	Voltaren	arthritis	1993	300
Glaxo	Zantac	ulcer	1995	1 000
ICI	Tenormin	high blood pressure	1993	410
J&J	Monistat	antifungal	1991	130
Lilly	Ceclor	antibiotic	1992	525
Marion Merrell Dow	Cardizem	angina pectoris	1992	500
	Seldane	antihistamine	1992	350
Merck	Sinemet	Parkinson's disease	1991	115
	Dolobid	arthritis	1992	100
	Flexeril	muscle spasms	1992	110
Pfizer	Procardia	angina pectoris	1991	233
	Feldene	arthritis	1992	140
	Glucotrol	oral antidiabetic	1994	305
SmithKline/Beckmann	Tagamet	ulcer	1994	515
Syntex	Naprosyn	arthritis	1993	560
Upjohn	Ansaid	arthritis	1993	560
	Xanax	tranquillizer	1991	370
	Micronase	oral antidiabetic	1994	200
W Lambert	Lopid	anticholesterol	1993	230

Source: Generic Pharmaceutical Industry Association and industry sources.

expired by the mid-1990s. The same trend is occurring in other industrialized markets and its effects are accentuated by changes in policy. Several governments, for example, are streamlining their procedures for approval of generics.[6] Others are going further by requiring that pharmacies supply patients with generic drugs unless a physician insists otherwise.

The world market for generics is estimated to have reached $15 billion in 1991. That figure is nearly double the level in 1986 but the pace of growth could even accelerate in the future (see Chapter 8). As it does so, more integrated firms will turn themselves into bifurcated or 'dual producers', supplying both generics and patented products.[7]

This will require several adjustments, however. Integrated firms are accustomed to selling their drugs through doctors, hospital administrators and owners of specialized clinics. In doing so they use traditional methods of marketing, known as 'pushing' their products down the distribution chain. If they are to become dual producers of patent-protected drugs and generics, they will have multiple points of sale and these traditional methods will have to be supplemented by efforts to 'pull' the products down the distribution chain. The latter can be accomplished by creating preferences among end-users – that is, the patients themselves.

Generics must also be sold in a manner which is quite different from that of on-patent drugs. Price is crucial and dual producers will encounter opposition from some of the larger companies specializing in these products (see Box 7.1). They must also learn how to make use of low-cost methods of distribution such as sales of large production runs to wholesale distributors. Opportunities to cut distribution costs should grow as new companies which specialize in the distribution of generic drugs emerge to handle these tasks on behalf of the larger producers.

By the end of this decade the industry will include a number of large firms producing both generics and branded drugs. The strategy, however, is not without drawbacks. In particular it can prove difficult

---

Box 7.1   Competition in markets for generics and OTCs

The move by integrated firms into generic markets poses a clear threat to producers that have traditionally specialized in these products. Most generic suppliers are too small to mount much of a challenge but some of the larger ones – for example, Mylan Laboratories, Bolar Pharmaceutical and Rugby Laboratories – are already beginning to react. Several are trying to strengthen their research departments in order to compete better with integrated producers. Others are forming alliances with research institutions that have new products or licences to sell.

There is less resistance in the case of OTCs, where large firms are moving rapidly to consolidate their position. Here they are motivated by growth forecasts that exceed those for the prescription market. That would be a reversal of the situation in the 1980s, when OTC sales grew by 6–7 per cent each year compared with 10 per cent for prescription drugs. Large integrated companies want to build up their OTC divisions to share in this growth and in hopes of extending the profitable life of their own prescription drugs.

to run a company with several sets of products and markets that differ widely. Some companies will be forced to focus on a single set of products, with varying degrees of in-house skills being deployed to minimize the distribution costs of generics and to maximize the effectiveness of R & D.

Dual production will not be the only type of marketing strategy to gain in popularity in the 1990s. Other types of speciality firms will also emerge in order to exploit opportunities that are not economically attractive to large, fully-integrated producers. One version would involve manufacturers and marketers of 'Lazarus drugs'. These are compounds which have been discarded after development because the potential market is too small.[8] These compounds could provide specialized companies with a modest product portfolio. Investors who have grown tired of the long wait for research programmes to become profitable could also have an interest in Lazarus drugs. The large firms which first developed these drugs can be expected to collaborate: they would have an opportunity to make money from compounds in which they invested research funds but from which they have realized no return. Finally the specialist in Lazarus drugs can minimize its risks by searching for niche products that can be marketed relatively inexpensively.

Small firms can find still other options. Some, for example, can obtain the rights to complete the testing process on compounds which were under development in the laboratories of a large company but were left unfinished. Others may acquire drugs which have already received government approval and then reformulate the products to improve their effectiveness.[9] Such tactics carry a risk: small speciality firms generally have sales in the range of $10 to $100 million and can soon find themselves in trouble if regulatory approval is slow in coming. Nevertheless, by picking drugs which are already in advanced stages of testing, the new companies acquire an accurate road map to the market, complete with extensive clinical data.

OTHER STRATEGIES

The way companies choose to organize their research and marketing activities will have implications for other strategic decisions. Examples include the firm's willingness to seek additional competitive advantages through mergers or acquisitions, to boost their research or marketing capabilities through more informal types of alliances, or to experiment with a variety of other tactical and strategic approaches. This section examines recent and likely trends in several of these areas.

Mergers and Acquisitions

The composition of the world's largest pharmaceutical companies was remarkably stable from the early 1960s until the end of the 1980s. Though individual firms have moved up or down in the ranking, there were almost no new entrants among the major producers and no firm claims more than 3–4 per cent of the world drug market. Suddenly, at the end of the 1980s, this long period of peaceful coexistence was disrupted. The number of mergers and acquisitions reached an all-time high, which unsettled scientists and managers alike. Estimates for the period 1988–90 put the total value of mergers and acquisitions at around $45 billion. The pace has now slowed but the industry has nevertheless entered a new phase where mergers and acquisitions will be more commonplace.

The data in Table 7.2 illustrate this trend, showing only some of the mergers and acquisitions that have occurred in recent years. Not only has the number of such agreements risen but so, too, has the size of the firms involved. The 1989 merger of two large American and British companies – SmithKline Beckman (SKB) and Beecham – marked a turning-point for the world's larger drug producers.[10] That merger was one of the largest in the pharmaceutical industry's history and underlines the apprehension which even the biggest firms face when planning their strategies for the 1990s.

The trend towards larger firm size seems to run counter to the tendency for greater specialization which was noted above. To some extent, that is true. Not all strategists accept the arguments in favour of greater specialization; some regard mergers and acquisitions as a way for firms to focus their activities on one phase of the industry, such as research, marketing or even brokering. Those that believe a larger scale of operation is essential find their justification in the changing circumstances which surround today's research and marketing activities. In the past, companies relied on patents to secure them a temporary monopoly. Even after the patents expired, brand names

Table 7.2   Mergers and takeovers in the drug industry since 1985

Company	Date	Combined prescription drug sales 1988 estimate (US$ bn)
Monsanto/G.D. Searle	1985	1.0
Eastman Kodak/Sterling	1988	0.8
SmithKline Beckman/Beecham	1989	5.4
Dow/Marion/Merrell	1989	1.9
APH/A.H. Robins	1989	3.1
Bristol-Myers/Squibb	1989	4.1
Fujisawa/Lyphomed	1989	1.4
Novo/Nordisk	1989	0.6
Mérieux/Connaught	1989	0.6
Rhône-Poulenc/Rorer	1990	2.8
Kabi/Pharmacia/Leo	1990	1.1
Roche/Genentech	1990	2.1
Sanofi-Sterling[a]	1991	4.7

Note
[a] Certain features do not confirm to the conventional form of merger (see Box 7.3).

Source: Scrip and industry sources.

could confer a significant measure of protection from competition. Under these conditions the producer of a new drug could usually expect real rates of return to average 9–10 per cent per year over a period of 15 years or more. By the beginning of the 1990s the combined effects of patent erosion, new research methodology, competition from generics and a rise in the real costs of R & D had changed these calculations. The profitable lifetime of a new drug was down to five years and its value was some 40 per cent less than in 1980.

It is the consequences of these developments that have forced many companies into mergers and acquisitions, but the implications for research and development are somewhat different. Even the largest firms concede that they cannot cover all areas of research and technology without outside help. However the idea that a further increase in the size of the firm will improve research performance and productivity is questionable. Those that endorse this view argue that a large integrated establishment should be able to field formidable teams in all or most of the promising research fields, while medium-sized companies must focus their research on selected areas.

Many analysts are prepared to concede that there are few advantages to be gained by building up larger research units (see Chapter 4). Innovation is thought to be most effective when carried out on a comparatively small scale. The underlying source of difficulties, they argue, is not that the research effort is selective but that research managers are not astute in selecting the compounds to be pushed through the research pipeline.

Nevertheless it is obvious that the task of research management is complicated. The high cost of today's clinical trials and animal tests forces companies to hold back research on promising compounds. Their hesitancy would not have mattered greatly in the past, so long as one new product proved to be a real success. Such delays are now risky because the industry's overall pace of innovation has quickened and the profitable lifetime of new drugs has fallen. The remedy which many companies have chosen is to develop a network of informal alliances (see next section) rather than opt for the more radical solution of a merger or acquisition.

The case for mergers and acquisitions is stronger when attention turns to marketing. Larger operating units are a logical way to cut marketing costs and gain access to new buyers. Few drug producers (particularly European ones) have distribution systems in place in both Japan and the United States, the world's two biggest markets. Many more firms lack experience in selling both OTC drugs and on-patent products. Either of these shortcomings could be remedied with the help of a skilful merger or acquisition.

Despite the marketing benefits, the criticism of recent merger results continues. Apart from the questionable effects on research, critics argue that drug producers could make better use of available funds by buying in capable staff rather than firms. The costs of such a strategy are also high and the risk of a mismatch between participants is great. Equally dangerous, many of the firms up for sale have something wrong with them (frequently the threat of liability claims). Finally the experience of the leading firms shows that most of the successful ones have relied on organic growth and this may be the soundest method to achieve the necessary economies of scale.

These economic arguments are convincing but even more powerful are the political drawbacks of consolidation. Today industries operate in a world where bigness is seen in a dubious light. The suggestion that giant oil multinationals or energy conglomerates could have unknowingly caused environmental disasters or suffered breakdowns in standards of safety and quality is simply not accepted by much of the public. The pharmaceutical industry is constantly subject to charges of conspiracy, price fixing or excessive profits. One way in which the industry deflects this criticism is to point out that there are thousands of drug producers and even the largest commands no more than 4 per cent of the world market. Attacks on the industry would only be heightened if, like the oil or automobile industry, eight or ten companies accounted for most of the world's drug sales.

Such dangers are obviously of less significance when attention turns to small or even medium-sized producers. Some of these firms use mergers and acquisitions to further their goal of specialization in certain market niches. That practice is popular among Scandinavian companies, which are relatively small by global standards (see Box 7.2).

Informal Alliances and Inter-firm Coalitions

A number of drug companies accept at least a portion of the logic behind the merger strategy but choose to implement it through informal agreements. Strategic alliances of this type go beyond normal market transactions but fall short of merger. Included are a whole variety of agreements such as the cross-licensing of research, joint ventures and joint companies, agreements on co-marketing and various other types of distributional arrangements that permit the originator of a new drug to gain access to additional markets.

The fact that most drug producers have traditionally been secretive and reluctant to cooperate with rivals has, until recently, limited the use of these informal coalitions. Now the tactics are gaining in popularity. The more cooperative attitude of today's firms is especially evident in the way they organize their R & D. Many of the larger drug producers are already accustomed to locating centres for R & D outside their country of origin. Research centres may be sited abroad

---

Box 7.2   Building specialization through mergers

The creation of Novo-Nordisk from two smaller Danish companies illustrates the way small firms can build up their areas of expertise through mergers. The new company, which was established in 1989, has a strong position in two niche markets: products for diabetes sufferers and industrial enzymes. Such a narrow product base is dangerous if either market should fail but the merger has also brought advantages. Research spending by Novo-Nordisk is still tiny by world standards but the merger provided the firm with ample research capability in its two areas of specialization. Novo-Nordisk has also gained the potential to realize economies of scale, since it is now the world's largest producer of industrial enzymes and the second biggest manufacturer of insulin. Finally the merger offers opportunities to rationalize distribution systems and to begin marketing products outside Europe.

---

to take advantage of lower costs or to be closer to acknowledged centres of research excellence. The popularity of this approach varies from country to country, though the number of examples is growing (see Box 7.3).

Large pharmaceutical companies may enter into informal marketing arrangements for several reasons. The most important is the steady rise in research spending and the fall in research productivity which has already been noted. Another is that consumers are becoming more involved in their own treatments, meaning that the markets for OTCs and diagnostic drugs are expanding very rapidly. It is essential that drug companies get their products into retail outlets and expand into new territories quickly in order to take advantage of this shift in buying patterns. The conversion of prescription drugs into OTCs represents an even more significant development in view of the many products coming off patent. This shifts the point of sale from doctors' offices to wholesalers and retail outlets, but the distribution skills needed to service these outlets are quite different.[11]

One example is the joint venture negotiated between Johnson & Johnson (J&J) and Merck Pharmaceutical Company. Established in 1989, the basis for the venture is that Merck's research skills complement J&J's marketing skills. Such practices are already well-established in the United States and becoming more common in Europe.

A slightly different approach has been taken by Proctor & Gamble (P&G). The firm spends heavily to develop its own products but its

---

Box 7.3   Strengthening research through informal alliances

Two of the biggest deals in 1990 were alliances or joint ventures rather than mergers. They were Sanofi's joint venture with Sterling Drug and Du Pont's with Merck. These alliances will offer a model for others in the 1990s.

The Sanofi–Sterling alliance provides an innovative solution to both companies' competitive and financial pressures. An important aspect is that the alliance entails no exchange of funds or goodwill depreciation. Certain development costs will be shared but basic research will remain independent. The alliance, however, will have exclusive rights to the research results of the two companies.

The Du Pont–Merck agreement is somewhat different. Sanofi and Sterling are comparable in size and research orientation. In contrast, Du Pont had spent heavily on research but lacked the development skills to obtain an adequate return, while Merck was encountering problems managing its research operation and faced a challenge to its research supremacy from Bristol-Myers Squibb. The alliance is partly a product swap, with Du Pont getting the development capacity it needs to finance and create an international sales operation. Meanwhile the newly created firm of Du Pont Merck Pharmaceuticals is expected to help Merck manage the growth of its research programme.

Other companies have found still more variants in order to link their research with academic or institutional groups. Examples include Hoechst's sponsorship of work being done at Massachusetts General Hospital, Merck's five-year agreement with Immulogic Pharmaceutical Corporation (begun in 1986) to develop drugs for auto-immune diseases such as rheumatoid arthritis and juvenile diabetes, Bristol-Myers Squibb's investments at Oxford and Strasbourg and Lilly's agreement with the National Cancer Institute of the United States on joint research into chemotherapeutic methods coupled with monoclonal antibodies.

---

entry into pharmaceuticals depends mainly on the establishment of multiple alliances with research-oriented partners. P&G's strengths are in advertising, marketing and distribution. Traditionally these skills were deployed to create dominant positions in various product markets, but, as the markets for OTC drugs have grown, the firm's focus has shifted. It has begun to use its marketing skills to reach retail chemists' shelves on behalf of other pharmaceutical firms. By 1988 P&G was one of the largest distributors of OTCs in the United States, with sales in excess of $500 million.[12]

Inter-firm agreements are not only pursued by large, research-oriented firms: they may also involve small firms (see Box 7.4) or may link up collaborators without any particular area of specialization. In the latter case, the prolonged time required for research and testing is

---

Box 7.4   Product swaps between large and small companies

One of the more popular types of alliances is the product swap between big and small companies. The smaller company with innovative research may wish to license out or co-develop the rights of a promising product. In return it receives the rights to market several older products with small markets. This enables the smaller company to fund future clinical and marketing development without seeing low-valuation equity offerings. The larger company simultaneously funds the research effort of the long-term product with the proviso that the smaller company retains the patent and receives royalties on sales if the drug is successfully marketed.

An example is Ciba-Geigy's $30 million equity investment in Isis, a small company based in California. The agreement, which also includes a five-year cooperative research investment, applies to the use of innovative techniques in molecular biotechnology.

---

frequently the rationale behind the alliance. Although a firm may have promising products in its research pipeline, several years of additional work may be required. During that time its sales force would be under-utilized. One solution is for the firm to acquire the marketing rights for other drugs in exchange for its collaborator's assistance in commercializing its own research.[13]

INDUSTRY STRATEGIES IN JAPAN: A UNIQUE CASE

Japan's pharmaceutical industry, like its manufacturers of automobiles, semiconductors and machine tools, has grown large and profitable while operating in a protected domestic market. Yet the country's drug producers have followed a line of development that sets them apart not just from their competitors but from most other domestic industries as well.

What are the features that make Japan's pharmaceutical companies unique? One is that the industry is uniquely domestic. Large drug companies in the United States and Western Europe derive at least a third of all their sales from foreign markets, while in some cases the figure exceeds three-quarters. In Japan this share is surprisingly low – only 6 to 8 per cent of total production. Another distinguishing characteristic is the industry's slowness to develop a research capability to match that of its rivals. Such moves have contributed much to the country's success in industries like automobiles and electronics, but Japanese-developed medicines, whether exported or sold through other drug firms, account for only a small portion of world markets – roughly 2 per cent of the $70 billion spent on drugs in the United States and Europe each year.

The explanation for this uniqueness is that the industry was nurtured by a branch of the government with a very distinct set of priorities. Most Japanese manufacturers are answerable to the Ministry of International Trade and Industry (MITI) but pharmaceutical companies are watched over by the Ministry of Health and Welfare (MHW). Drug producers have therefore been guided by a national health policy, with the primary goal being a stable supply of modern drugs. The MHW worked to achieve this goal by limiting foreign competition but did not restrict the use of foreign drugs. Many medicines were produced under licence and eventually the Japanese market came to be dominated by domestic formulations of drugs designed in other industrialized countries. Meanwhile the industry's profitability was sustained by an equally unique set of policies governing methods of drug prescription, reimbursement and other types of support.

This combination of policies is now being scrapped. The MHW is drastically cutting the prices of many drugs (see Chapter 6) and has removed restrictions on the entry of foreign multinationals. Rather than simply licensing their products to Japanese firms, foreign companies are now permitted to buy into domestic firms (often distributorships). A number of multinationals have quickly gained the right to develop, manufacture and distribute their medicines domestically. By 1990 foreign firms had captured 21 per cent of the vast Japanese market.

These changes in policy have forced domestically-owned firms to alter their own strategies. So long as drug companies operated in an insulated and highly profitable domestic market they were complacent and had little incentive to export. Nor were there strong reasons for the Japanese to embark on ambitious research programmes since the latest drugs could easily be obtained through licensing. Cuts in domestic drug prices and the entry of foreign competitors have reduced the profitability of the home market and increased the appeal of exports.

Taking these developments as a basis, several analysts have predicted that Japanese drug firms, like their counterparts in the automobile and electronics industries, would soon become exporting power-houses. The export drive has been slow to materialize, primarily because drug companies' research programmes were limited and their emphasis on selling imported drugs left them in poor shape to compete internationally.

Japanese producers are gradually beginning to adapt to the new conditions. Initially they relied on licensing agreements to get promising products into overseas markets.[14] Foreign partners with established sales forces in Europe and the United States marketed these drugs in return for a large share of total revenues (sometimes as much as 70 per cent). The licensing phase of the Japanese industry now seems to have come to an end. All the larger companies – Takeda, Sankyo, Tanabe, Fujisawa and Yamanouchi – are seeking more tangible and profitable ways of operating in foreign markets.

Their entry into international markets poses several problems, however. First, the largest companies (Takeda and Sankyo) are far smaller than their foreign rivals and will face intense competition in overseas markets. The biggest firms in the United States and Western Europe are themselves being forced to merge in order to survive as drug markets become more competitive. Second, Japanese firms are not especially profitable. In 1990 the average gross margin of the country's ten largest companies was 55 per cent – substantially less than that of the leading firms in the United States. Few Japanese firms can yet afford to launch an expensive overseas programme, through either exports or FDI. Third, the hurdles faced by would-be exporters involve both distribution and research.

In order to be effective, Japanese exporters will require a sizeable overseas sales force (up to 2000 people for markets in the United States or Europe) but this is expensive and time-consuming. At the beginning of 1991 only two products were being marketed in the United States and Europe by their Japanese originators. Meanwhile the industry must strengthen its research effort. Most of the drugs which have been developed by Japanese firms represent minor changes in well-known chemical structures. Real export potential depends on the design of new drugs that provide completely new cures.

The Japanese response to these problems is to rely on FDI, buying up existing research and marketing operations in Western Europe and the United States, but that tactic has drawbacks. Most firms are too small to purchase a large European or American rival and many

*Table 7.3   Recent links initiated by Japanese drugs groups in the
              United States and Europe*

Year	Purchase/link
1988	Fujisawa buys majority stake in Kling Pharma, a German drugs company, for £33m
1988	Sumitomo invests $10m in Regeneron (United States), a drugs company researching neurological products
1989	Yamanouchi buys Shaklee (United States), a vitamin and healthcare company, for $395m
1989	Eisai sets up a European sales and marketing base in London
1989	Chugai buys a stake in British Bio-technology (United Kingdom), a biotech company (£3m)
1989	Eisai opens a drug research unit in Massachusetts
1989	Fujisawa buys LyphoMed (United States), a drugs company, for $750m
1989	Dainippon signs a joint venture with Rhône-Poulenc (France) on marketing Dainippon products in Europe
1989	Chugai buys Gen-Probe (United States), a diagnostics company, for $100m
1989	Daichi agrees on development project with Chiron (United States), a biotech company
1989	Chugai agrees on partnership with Rhône-Poulenc (France) on biotechnology drug
1989	Japanese drug companies discuss an informal Europe-based association to spearhead moves into Europe
1990	Yamanouchi signs licensing agreement with Microgenics (United States), a company in medical diagnostics
1990	Sankyo buys a majority stake in Luitpold-Werk, a German drugs company
1990	Takeda sets up joint ventures in Germany, Italy, France and the United States
1990	Yamanouchi buys the drug division of Gist-Brocades (Holland) for $290m
1991	Meiji Seika buys 60 per cent of Tedec, a Spanish drug firm

Source: *Financial Times*, 22 January 1990.

successful purchases are certain to encounter political opposition. The industry has nevertheless begun to move rapidly. The information in Table 7.3 summarizes the results, showing purchases, joint ventures and other initiatives concluded in the period between 1988 and early 1991. Some of these moves are intended to build up distribution channels. The types of links being sought are usually with firms which have little research expertise but know how to manufacture a drug cheaply and sell it quickly. Such acquisitions may be less profitable than the purchase of larger firms, but they are also less politically sensitive and less expensive. In the case of research the approach is two-pronged. Japanese companies are trying to link up with foreign research teams in order to gain insights in fields such as biotechnology and medical diagnostics. Meanwhile their domestic research programme has been reoriented in order to develop drugs that are both profitable and have export potential (see Box 7.5).

In conclusion the Japanese industry is in the midst of a significant transformation. Its moves overseas are complemented by a substantial

---

Box 7.5   Research stages in the Japanese pharmaceutical industry

The industry moved quickly to develop its own research capability in antibiotics. These were originally used to treat the bacterial infections that plagued post-war Japan. The industry's growth during the 1980s relied heavily on the research results of the previous decade. It depended primarily on 16 proprietary drugs which were developed by Japanese firms. The fact that half of these drugs were antibiotics proved to be a serious drawback, however. Such drugs have especially short product lives and are regarded by some analysts as the 'low-tech end' of the industry.

A second stage in the evolution of research prowess was to design medicines to treat the illnesses of the rich and elderly. Today most of the drugs sold on the Japanese market were invented in that country. The majority, however, are not completely new ideas but modifications to chemical structures that were first designed in other countries.

The third stage in the research programme which is currently under way is strongly supported by policy makers. Drug prices have been reduced by 45 per cent over the period 1982–90. Companies, however, can negotiate much higher prices for new drugs, which ensure that they are profitable for a number of years. Such a pricing system is proving to be a powerful spur to innovation. Meanwhile the industry has redirected its research effort. Most of the medicines under development are intended to treat modern problems, such as cancer, thrombosis, osteoporosis, senile dementia and the rejection of organs.

upgrading of domestic research on products aimed at the international market. However the industry will not be quickly converted into the large-scale exporter that some had first expected. Instead it is more likely to rely on its rapidly expanding network of formal and informal links with foreign firms, moving straight into a system of offshore research and marketing that complements its domestic operations.

INDUSTRY STRATEGIES IN DEVELOPING COUNTRIES

The many hurdles, delays and problems encountered by the Japanese are instructive when attention turns to developing countries. If Japan, a country with a huge domestic market, ample financial resources and abundant technical skills, requires such a long time to build up a competitive pharmaceutical industry, what are the strategic options available to firms in developing countries?

The choices are comparatively few, primarily because producers in developing countries lack research expertise. China seems to be one exception. That country's growing links with firms in industrialized countries reflect the latter's interest in developing retail and OTC products from traditional Chinese herbal remedies.[15] Some foreign participants hope to make use of Chinese research on biological agents which they would convert into drugs. Other arrangements are more complicated, calling for Chinese scientists to conduct the research and for the collaborator to develop the products. Later the Chinese are expected to test the prototypes in research facilities provided by their overseas partner.

Few firms based in developing countries have such a research option. Only a handful have actually discovered a NCE and the instances of collaboration between pairs of firms that are owned and operated from developing countries are few.[16] A multinational usually initiates such contacts; the objective is invariably to gain market access. The multinationals' early moves into developing countries allowed little or no role for local industry. The growth of domestic demand and the lure of larger public-health budgets has made the markets of at least a few developing countries more attractive investment sites. That is the case in several Asian countries, where demand for medicines is growing by more than 10 per cent each year.

Local firms in developing countries which do not enforce patents on products have other options. They may select an existing NCE and then elaborate a synthesis for manufacture and sale in the domestic market. Such a tactic offers several advantages: the NCE can be produced in small batches in multipurpose plants, investment in R & D and production is low and the firm is able to supply a modern drug at highly competitive prices. Several minimum conditions have to be satisfied, however. The local firm must be capable of manufacturing medicinal chemicals. It must also be capable of conducting reproductive research. The latter skills are necessary in order to elaborate the analytical chemical specifications and to demonstrate the stability and bioequivalence which ensure that the local firm's version is interchangeable with the original product.

Most firms which adopt such a tactic will compete against imports and do not attempt to sell abroad. In a few instances, however, drug makers have opted for a more independent approach, using their production capacity and experience gained in their home market to launch export drives. No set of producers have pursued this line more actively than India's drug makers. Although the country accounts for only a tiny amount of the world market, its exports are rising fast.

Like many other developing countries, India recognizes patents only for processes. Some of its companies specialize in developing new chemical processes yielding drugs that are identical to those produced (at much greater expense) by multinationals. The Indian firms begin by selling their new products domestically; later they scale up operations to cut manufacturing costs. The foreign markets which they select as targets are countries that recognize process patents but not patents on the product itself. By the time the companies are ready to export they can often sell the drugs at less than a tenth the price charged by competitors in industrialized countries.

The export zeal demonstrated by Indian firms is only partly due to the country's patent system (see Box 7.6). It is this issue, however, which has figured most prominently in international disputes which have pitted India against multinationals and government officials in industrialized countries. These arguments could intensify in the future. The growth of generic markets and the many drugs which will soon come off patent mean that a greater portion of the world's markets could soon be open to producers in developing countries. The industrialized countries' concerns may eventually lead to some modi-

---

*Box 7.6		Policies and conditions contributing to India's export drive*

India's exports of bulk drugs were to the value of 4 billion rupees ($280 million) in 1989, more than double the level in the previous year. The country's patent system is only one of the reasons for exports having become so important to Indian drug makers. The shift is also due to a steep decline in profit margins. According to the industry, the average profit margin was 15 per cent of sales in 1970 but had fallen to 4.5 per cent by 1988. That is far below the return allowed by the Drug Prices Control Order (DPCO). Profit margins are probably lower now than in 1988 since prices have remained static while the costs of labour, raw materials and power have escalated.

Stringent price controls have ensured that Indian drug prices are among the lowest in the world – an important consideration since almost half the country's population is below the poverty line. Today a packet of ten aspirins sells for as little as ten rupees ($0.62). It is this factor which explains why Indian drug companies have turned to different strategies. A favourite practice has been to shift production from the domestic to the foreign market. Exporters obtain better prices for their bulk drugs, they are eligible for export incentives and claim income tax exemptions for earning scarce foreign exchange.

Other strategies are also being used. Several companies are beginning to specialize in the production of OTC formulations, since these can be sold outside the jurisdiction of the DPCO. A more disturbing development is that many pharmaceutical companies are moving into other fields, such as basic chemicals or dyestuffs, or are withdrawing from the drug industry entirely. The danger is that the industry may not be able to meet overall demand – because of a fall in investment and a shift to export markets.

---

fications in the patent laws of developing countries, but the latter are unlikely to accept a 20-year patent on new drugs.

Policies governing price controls and import restrictions lead to other types of strategies. Some developing countries have no price regulations on new products although price changes are strictly controlled. In such situations local firms try to obtain maximum price flexibility by periodically submitting a flood of applications for new launches which are actually only slightly modified versions of the original products. A similar tactic is followed when the import price is used as a benchmark for setting the price of a domestically-produced version. Firms first import small quantities of the drugs at artificially high prices. Once the benchmarks are set, they take out a licence and produce the drugs locally. These practices explain why several thousand products are sometimes registered in a single year.

Various types of aid and foreign assistance can reinforce or complicate the strategies of drug producers in developing countries. Some of this aid comes as donations and grants. More often it takes the form of credit lines which are a direct gift of pharmaceuticals or are tied to purchases from suppliers in the donor country. Grants or donations can supplement health budgets and provide a boost to local manufacturers. The home industry, however, often regards tied funds or direct gifts of drugs as unfair competition and may oppose assistance in these forms.

The bulk of pharmaceuticals are distributed through publicly-controlled channels and this practice also has other strategic implications.[17] Essential drugs are usually obtained by competition through open tenders. Free competition is possible because either the patents on most essential drugs have expired or patent protection is not demanded for public health purchases. The relatively large volumes involved attract many bidders, particularly domestic producers of generic drugs.

Open tenders are a cost-effective way of obtaining essential drugs but they can also pose problems for local firms. First, the lowest bid should always win, though governments frequently negotiate in order to obtain the best deal for visible costs at the expense of hidden costs in terms of poor quality or lengthy delivery times. Second, profits are small and the uncertainty of the process pushes up costs. Nearly all essential drugs are multisource products, which makes price competition very intense, while hospital packings reduce profitability further. Winners may change from year to year, meaning that annual programmes cannot be planned. Either capital is tied down in slowly-moving inventories or the firms' delivery times are not competitive. Third, foreign competition for tenders is intense and the local industry is frequently unable to approximate winning prices. Instead quotations are based on marginal costs, with the intention of keeping the plant running rather than generating profits. For these reasons, the tender business attracts only those manufacturers which also produce essential drugs for the private market. The production of additional quantities for tender contracts is then more feasible both from technical and economic points of view.

To sum up, the strategies available to firms in developing countries are meagre, but the industry is changing rapidly. Some of these developments may offer new options, though they will also bring additional problems. The concluding chapter of this book looks at the prospects for the pharmaceutical industries of industrialized and developing countries.

NOTES

1. The experience of SKB, once a full-line R & D based company which faced particularly strong competition from generics, is typical. The company was unable to bring any new drugs to the market at the time when the patent life of its major money-earner was expiring. SKB's net margins fell regularly from 1984 until it merged with Beecham in 1989.
2. An agreement between Johnson & Johnson and a British firm illustrates this tactic. The collaborator undertook to find buyers for 300 patents from the product portfolio of Johnson & Johnson. In return the British firm receives 50 per cent of the income generated.
3. A particular strategy may be very attractive for any individual firm, but if a great number of competitors adopt the same tactics, the strategy's own popularity becomes self-defeating. This possibility is known to economists as the 'fallacy of composition'.
4. An example is the pain-killer Advil, produced by American Home Products Corporation. The drug continued to be profitable even when sold through OTC channels. In 1988, Advil accounted for over half the OTC market in its particular class of pain-killer. In the United States the market for all such drugs was $2.2 billion in 1989 and Advil's share was 11 per cent.
5. Another consideration has to do, not with research, but with demand. The United States Medicare Catastrophic Coverage Act requires that all Medicare patients be prescribed generic drugs. World-wide, the market for generics is expected to reach $15 billion in 1991.
6. This change was part of the Waxman/Hatch Act. Now the generic companies need to prove only two attributes. They must show that their product's active ingredient is chemically identical to that of the branded drug and that it is absorbed into the body to the same degree as the branded drug.
7. Firms in the United States which are developing a dual line of business in generics and patented drugs include Warner-Lambert, Lederle and Wyeth Laboratories. In Europe, Fisons, Ciba-Geigy, Rhône-Poulenc and Hoechst also have subsidiaries making generics. The merger of some of Rhône-Poulenc's operations with those of the Rorer Group in 1990 is part of a larger strategy to make the former company Europe's largest producer of OTC drugs.
8. Drugs with potential sales of $10 to $50 million or with markets of only 20 000 to 30 000 patients are not regarded as worth being handled by integrated producers with very large sales forces.
9. A drug's delivery system, for example, may restrict its potential. If a drug must be injected several times a day, its market is likely to be limited. Reformulating drugs so that they can be administered more safely and efficiently –

for example, through skin patches or time-release pills – will boost demand and make a profit for the speciality firm.

10. The reasons for this merger were twofold. First, sales of SKB's prescription drugs were slowing in all markets, but particularly in the United States. Second, Beecham wished to continue its work on the treatment of chronic diseases where longer and more complex clinical investigations require increased research expenditures.

11. Among the drugs which have successfully made this switch are two pain-killers, Advil and Nuprin.

12. Examples are P&G's agreement to market drugs for Syntex and Gist-Brocades.

13. An example of such an arrangement is a recent agreement between Du Pont Company and Merck in which the latter exchanged the North American and European marketing rights for specific drugs for the rights to commercialize Du Pont's work in the field of heart drugs.

14. Cardizem, Pepcid and Cefobid are drugs developed by Japanese firms but buyers in the United States see them under the labels of Marion, Merck and Pfizer. Mevalotin, an anti-cholesterol drug, is expected to be the most successful, with international sales forecast to be $1 billion a year. Sankyo developed the drug but, because of its limited experience in foreign markets, licensed it to Bristol-Myers Squibb. Mevalotin was launched in the United States in March 1991.

15. The potential of traditional Chinese methods is suggested by the fact that the pharmaceutical industry's exports rose nearly tenfold between 1975 and 1987, amounting to $303 million in the latter year.

16. Where such cases exist a firm in a developed country may often have a minority interest in the originator of the initiative. An example is Unilab, which is based in the Philippines but has operations in Nigeria and South East Asia.

17. Tenders are formally open to all suppliers that meet the technical requirements but registration and GMP standards are frequently waived for state-run firms and small, local manufacturers. Public procurement practices may also allow different levels of price preference for local suppliers. This practice can be a useful and necessary form of support when a firm is starting up, but will distort price competition if continued over extended periods of time.

8.   A Postscript on Prospects

The foregoing chapters leave little doubt that the pharmaceutical industry is in the midst of a major restructuring. This transformation is proceeding most rapidly in industrialized countries, though firms in many developing countries are also having to make some adjustments and more can be expected in the future.

The forces behind this upheaval are many and will affect various parts of the industry in distinct ways. The outlook is further clouded by the types of policies to be introduced by governments in industrialized and developing countries during the course of this decade. The net effect of all these circumstances is to create a great deal of uncertainty, making it futile to try to construct any industry-wide set of forecasts. Several of the developments examined in earlier chapters will have a major impact, however, and their implications are sketched in below.

DEMOGRAPHIC TRENDS AND THEIR CONSEQUENCES

Pharmaceutical markets are being reshaped by two broad types of demographic trends – rates of population growth and changes in the age structure. Most countries fall clearly into either of two groups, depending on the overall demographic pattern. In one group are countries where the total population is constant or declining while the median age is rising. In another group are countries experiencing a rapid increase in population, although the median age is low and is not increasing.

Table 8.1 illustrates the effects of these demographic shifts for a sample of industrialized and developing countries. The fact that the population of many industrialized countries is stagnating or declining is no surprise, although the rapidity of the aging process is remarkable. One-sixth of all Germans will be over 65 by the year 2000 and

Table 8.1   Demographic trends in selected countries, 1990–2010

	Annual growth (%) 1990/2010	Share of age group <15 in total population		Share of age group >45 in total population		Median age (years)	
		1990	2010	1990	2010	1990	2010
Industrialized countries							
Belgium	+0.1	18.3	17.3	37.0	44.6	36	41
Canada	+0.7	21.3	18.6	30.9	41.8	33	39
Finland	+0.1	19.2	16.4	34.9	45.2	36	42
France	+0.2	20.3	18.0	34.4	42.4	35	40
Germany	-0.3	15.2	14.5	41.4	49.7	39	45
Italy	+0.1	17.5	17.5	38.3	44.2	37	41
Japan	+0.4	19.2	18.9	36.7	43.5	37	40
Netherlands	+0.1	17.9	14.9	33.8	46.3	35	43
United States	+0.7	22.1	20.6	30.9	39.5	33	37
Developing countries							
Argentina	+1.2	31.1	26.4	26.9	28.7	27	29
Brazil	+1.6	35.2	28.2	17.5	24.7	23	27
Chile	+1.2	28.8	24.5	21.7	30.6	26	31
China[a]	+0.9	25.4	21.4	21.4	32.1	26	35
Colombia	+1.6	36.2	28.7	15.8	23.4	22	27
Egypt	+1.8	39.2	29.6	15.6	21.8	21	26
India	+1.4	34.5	25.7	18.8	25.2	23	29
Kenya	+4.0	52.6	48.5	9.2	9.6	14	16
Mexico	+1.8	39.1	29.4	14.1	20.7	20	26
Nigeria	+3.5	48.6	47.2	11.6	11.1	16	16
Pakistan	+2.1	43.0	34.2	12.6	16.6	18	23
Republic of Korea	+1.1	32.4	24.8	19.8	30.7	25	32
Saudi Arabia	+3.2	44.8	40.2	12.4	15.5	18	20
Zaire	+3.1	45.3	43.7	13.3	13.5	17	18

Note
[a] Includes Taiwan Province.

Source: International Labour Organization, *Economically Active Population, 1950–2025*.

by 2010 half the country's population will be 50 years or older. The populations of Canada, Italy and Japan are younger but aging rapidly.

Surprisingly the aging process will be fastest in a number of developing countries. The share of the elderly and those over 45 will not rise significantly but the absolute numbers will soon exceed the totals for most industrialized countries. For example, by 2010 the number of Chinese over 65 will be as large as that in Western Europe in the year 2000. At the opposite end of the spectrum are developing countries which fall into the second group. Examples are Kenya, Nigeria and Zaire. In many of these countries the median age in 2010 will be less than half that of industrialized countries and in some cases below 20 years.

These demographic trends are altering important features of drug markets. The frequency of drug consumption is rising most rapidly in markets where the population is aging. At the same time the pattern of consumption is changing. Chronic and degenerative diseases are prevalent after an age of 50 and dominant among those over 60 years. The markets for medicines to treat these diseases are therefore growing.[1] The cost of drugs favoured by different types of specialists reinforces these effects. Neurologists, for example, prescribe the most expensive drugs, while paediatricians opt for the cheapest ones.

The effects are different in the second group of countries, where the population is young and increasing. Young people are subject to acute or infectious ailments. The need for drugs is particularly great among those under five years as a result of vaccination programmes against childhood diseases, prevention of vitamin and mineral deficiencies and treatment of common colds and associated infections. The unfavourable economic and social conditions which prevail in many poor countries mean that the need for such drugs is particularly great in this age group.

The frequency of drug consumption in poor countries also appears to be falling, although this cannot be attributed to any improvements in health or the relative young age of the populations. Per capita consumption of drugs in most of these countries is already very low and the prospects for growth in income or improvements in public health care are bleak. Because of these constraints very little of the population increases in these countries will be translated into greater demand for pharmaceuticals.

The demographic shifts described here are already changing the way governments influence the pattern of consumption. As much as two-thirds of all drugs purchased in developing countries are obtained through private rather than public channels. That figure – which reflects the inadequate systems of public support – is high but will grow as populations increase. Meanwhile regulators in industrialized countries are searching for ways to contain the costs of public health care and have chosen to focus on reimbursement programmes. The reasons for this decision are political as well as economic: spending on drugs accounts for 5 to 15 per cent of the health care budget in industrialized countries but the industry's tarnished image makes it an appealing target for cost-cutters.

The new parsimony of industrialized countries will definitely slow the growth of consumption, though the long-term effects may not be as severe as drug companies fear. Health officials will soon have to search for more substantial savings in other parts of their budgets. Eventually the industry could even see its share of public health care spending rise, although the total amounts allocated for reimbursement programmes will decline. The reason is that drugs tend to be cheaper than hospitalization and policy makers could encourage their use as a first line of defence.

How will these demographic trends and associated policy changes affect global markets? There is a general consensus that the demand for drugs will grow rapidly. Some analysts go on to suggest that the share of developing countries in world markets could also rise. They base their argument on the spectacular increases in the populations of these countries and the possibility that sales in industrialized countries will decelerate as generics and OTCs claim a greater market share and as reimbursement programmes are scaled back. A greater significance is attached to these trends than to counter ones, such as the acceleration of the aging process in industrialized countries and their generally better prospects for growth of income.

Figure 8.1 provides a set of historical figures which cast some doubt on this line of reasoning. Per capita sales of pharmaceuticals rose steadily from 1975 until 1984 in both industrialized and developing countries. Since then, there has been a departure from the long-term trend; per capita sales in rich countries have increased disproportionately. When these figures are translated into shares, the developing countries are seen to account for 20 to 23 per cent of world sales in each year during 1975–84. Their share has steadily fallen in later years, however. By 1990 less than 15 per cent of all pharmaceuticals were sold in developing countries, with the remainder being purchased by consumers in the industrialized world (see Table A.17).[2]

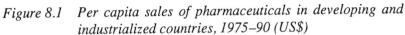

Figure 8.1   Per capita sales of pharmaceuticals in developing and industrialized countries, 1975–90 (US$)

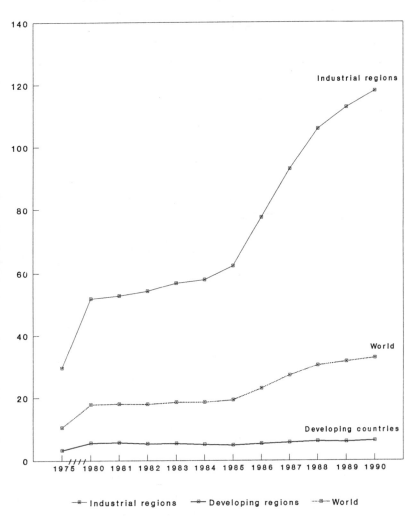

Source: UNIDO.

The consumption gap between the two groups of countries has been widening for at least five years and present trends are unlikely to be reversed significantly in the 1990s. The conclusion which emerges from this evidence is that population increases, however rapid, do not by themselves lead to a substantial growth in the markets for pharmaceuticals. Other factors such as income growth, the increasingly high priority which consumers in rich countries attach to personal health, and the greater number of elderly consumers are probably more important determinants. Other forces – in particular, policy changes and industry restructuring – are also at work and are discussed below. Their effects, however, are slower to emerge and patterns of consumption will take even longer to adjust.

POLICY OPTIONS IN THE 1990S

No radical departures from the existing policy framework are expected in the 1990s. There will, however, be a rearrangement of priorities. Governments in industrialized and developing countries alike will be forced to accept efficiency as a major goal of health policy.

The options available to policy makers are circumscribed to some extent by the industry's own performance and the types of products it has (or will have) in the market. Figure 8.2 gives a stylized view of competition over the marketable lifetime of a product which helps to illustrate this fact. When the new product is launched, it is a single-source drug which is protected by patent. Rival products – molecular modifications that are also patentable – are subsequently launched and the market share of the pioneer drug falls. The loss is gradual if the new products are simple 'me toos' which offer no therapeutic advantage; in that case the path depicted by the line ABDE shows the decline in the original product's share. Losses are more substantial if the new rivals are innovative, improved versions. Line ABCE then traces the development of the pioneer drug's market share.

Figure 8.2 shows that three forms of competition occur during the original product's lifetime. In the first stage the pioneer drug has no competitors and claims the entire market. Patent protection exists throughout the period AE, although the original drug still faces therapeutic competition from other innovators. Once the patent expires, competitors can legally use the same active ingredient as the originating firm and generic (multisource) competition takes hold.

According to this stylistic picture, policy makers have two broad sets of options. The more familiar is generic substitution, which con-

Figure 8.2   Monopoly and competition in the life cycle of a pioneer drug

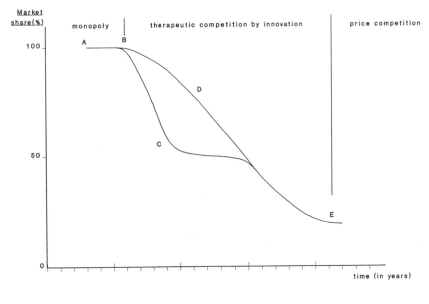

Source: UNIDO.

sists of measures intended to bolster competition once the original product's patent has expired. The approach is appealing to health officials since the provision of cheaper (but equivalent) drugs is an obvious way to demonstrate the cost-effectiveness of decisions. Generic substitution can also be implemented in a rather straightforward manner. The same does not apply to the alternative, which is known as therapeutic substitution. The choice of policies to encourage competition prior to expiry of the patent on the original product poses several medical and technical problems which have yet to be resolved. Therapeutic substitution is also opposed by the industry, since it would cause investment returns to begin to decline at a much earlier stage in the original product's life cycle.

Neither the developing nor the industrialized countries have yet to introduce a comprehensive mix of policies and measures that would lead to generic substitution. Brand-name loyalties, fragmented systems of distribution and the prominence of multinationals are formidable

Table 8.2   Policy options for generic substitution

Renegotiation of effective patent life	A reduction (possibly to a period of 10–12 years) would substantially increase generic opportunities
Official encouragement of generic prescriptions	Requirements may be mandatory or voluntary
Greater use of formularies	These are 'positive' lists of the reimbursable drugs for use mainly in hospitals and public health care institutions
New incentives for pharmacies	Regulators may set the price to be paid by the public health service and then add a profit; if the pharmacist buys at a lower price, profits are increased
	Alternatively regulators can ensure that the pharmacist is 'economically neutral' by allowing him a fixed fee, regardless of whether he sells the original or a generic
Extensive adjustments in reference price systems	Adjustments based on the availability of multisource drugs at different prices; so far, these moves have been limited to products with the same active ingredients
Critical monitoring of doctors' prescription costs	The objective is to force doctors to prescribe the cheapest available multisource drugs
Restrictions on promotion material	Require that material include the international non-proprietary name (INN) and information on the cost of treatment

Source: UNIDO.

barriers for developing countries, while the industrialized countries face opposition from the industry and the medical profession.[3] Policy makers will nevertheless adopt a more aggressive attitude in the 1990s. The choices available to them are many and some of the more likely ones are summarized in Table 8.2. They entail changes in patent laws and reimbursement programmes, enforcement of new regulations for the rational use of drugs and the introduction of new measures to influence prescribers and distributors.[4]

So far, generics have made their greatest inroads in the United States. Brand-name loyalty is not particularly strong in that country and consumers bear more than half the costs of the medicines their doctors prescribe. New laws that support the development of generics are therefore popular. Policy changes in other industrialized countries will gradually lead to a similar shift in consumption patterns. Western Europe is generally thought to offer the best growth prospects for generics in the future (see Box 8.1). There are differences between northern and southern Europe, however. The outlook in northern European markets is bright, mainly because patent protection is strong and prices are high. The situation is different in southern Europe where prices are low and patent protection is weak, which means that copying can be more profitable.

The opportunities for policy makers to pursue an aggressive policy of generic substitution beyond the mid-1990s are less certain. These products are part of a continuum in a long-term process of innovation and technological change. Conditions for generic competition are most favourable when a period of rapid innovation is followed by one of fewer introductions. The number of new drugs increased most rapidly in the period 1961–74 (see Table 4.1), meaning that the outlook for generic competition should be good until the latter part of the 1990s. The fact that innovation slowed in the 1980s could make it more difficult to enforce the policy in later years, however.

By the end of the 1990s health officials may be forced to place a higher priority on therapeutic substitution. Two alternative versions were depicted in Figure 8.2. Competition may take the form of molecular modifications and 'me too' drugs, but a more desirable outcome, when possible, would be competition between different chemical categories of drugs. Each of the rival products would be therapeutically distinct in terms of their chemical structures and mechanisms of action, although any could be used to treat the same disease.[5]

Box 8.1   Prospects for generic markets in industrialized countries

The size of the generic market depends on national policies and consumer preferences. In Europe the industry and the medical profession have combined to resist legislation to promote generics. Their strongest argument is that generics are unreliable, varying widely in terms of their potency. Box table 8.1 gives market shares and forecasts for a number of industrialized countries. Because generics are much cheaper than brand-name drugs, estimates for both volume and value are stated whenever possible.

Box table 8.1   Market shares for generic drugs, 1990–5

Belgium	Apart from hospitals, generics are of little importance
Denmark	Generics account for 50 per cent of the retail market and growth in the 1990s will be moderate
France	Generics are mainly sold in hospitals; they have 1 per cent of market value and three per cent of volume; their share should read 13 per cent by 1993
Germany	Generics lost market share with the introduction of fixed-price support systems for reimbursements; producers of originals cut prices to the new levels, losing revenues but expanding their market shares; generics at present account for 17 per cent of the market by value and 22 per cent by volume; they should have 35 per cent of the market by 1993
Greece/Ireland	There is no generic competition
Italy	Product patents were introduced in 1978 and will soon begin to expire; generic competition could follow
Netherlands	Generics have 16 per cent of market value and 30 per cent of volume; their share should reach 40 per cent by 1993
Portugal/Spain	The generic market is very small
United Kingdom	Generics account for 10 per cent of sales and 30 per cent of volume; their share should be 50 per cent by 1993
United States	Generics accounted for 18 per cent of prescription sales in 1990; demand is expected to grow by 14 per cent annually, giving these products a quarter of the market in 1995

Source: UNIDO estimates, Scrip, *World Pharmaceutical News* (various issues) and national producer associations.

An effective programme of therapeutic substitution will be complicated to implement, however. It must be based on a detailed assessment of the therapeutic improvements which each new product offers and the overall implications of such a programme for public health. These admittedly vague ideas are usually expressed in terms of various product categories, such as:

- major therapeutic advances,
- significant therapeutic advances,
- modest improvements in therapeutic abilities,
- minor therapeutic improvements,
- no improvement but a favourable regulatory opinion (usually as the result of a cheaper cost of treatment),
- unfavourable regulatory decision.

The scientific basis for such a categorization is limited. Moreover the exercise requires agreement between health officials, other government regulators, pharmaceutical companies, the medical profession and representatives of the consuming public.

Nor are these the only issues to be resolved. Governments must also determine the total number of therapeutic competitors which are 'desirable' in each product market. Severe restrictions on the number of competitors would negate a policy of therapeutic substitution, while an unnecessarily large number of competitors – many of which would be 'me toos' – can waste resources and confuse both distributors and customers (see Box 8.2).

Box 8.2   The dangers of product proliferation

The same active ingredient is often available under a multitude of different brand names. In Europe, captopril is obtained under 19 different brand names and enalapril is found under 25. Ranitidine can be found under 43 different names, of which 14 are marketed in Greece. The situation is similar for cimetidine which is available under 60 names (including those with the INN and company name) of which 22 are found in Spain. In extreme cases the situation becomes confusing for the doctor, the patient and even the pharmacist. An unnecessarily large number of branded products with the same active ingredient does not foster price competition.

Most industry representatives argue that the registration of a NME should depend only on medical criteria (efficacy, safety and quality) and that these decisions should be distinct from negotiations on pricing and reimbursements. However, if a new drug promises no overall improvement in therapy, the producer should at least demonstrate that it offers a saving for the customer. The need to demonstrate a saving seems to be particularly important in the case of chemically related 'me toos'. The same argument may be made for 'me too' drugs which have different chemical structures and offer no overall improvement in treatment.

PROSPECTS FOR POLICY HARMONIZATION

While countries struggle to put together programmes for generic and therapeutic substitution, the EC is moving towards its own goal of a single market. The industry is especially concerned about the effects of market integration on drug prices and reimbursement practices. The Community's efforts at policy harmonization are likely to be of wider significance, however, since they could serve as a model for similar initiatives on a wider scale.

The harmonization of policies on drug testing is one of the practical options being considered by the European Commission and which, if successful, could find application in other parts of the world (see Box 8.3). Such a move promises clear benefits for the industry since it

---

Box 8.3   International standards for animal toxicity studies

No estimates are yet available for the savings to be derived from standardizing all drug licensing requirements. Scientists at the Centre for Medicines Research in the United Kingdom (which is funded by the pharmaceutical industry) have analysed one particular issue: the duration of animal toxicity studies. They concluded that, apart from carcinogenicity testing, there was no safety advantage in continuing these tests for more than six months. The EC has adopted a six-month duration for toxicity tests, although the United States and Japan insist that animal tests continue for at least 12 months. The Centre estimates that, if the rest of the world were to adopt the EC's approach for this one requirement, the industry would save $100 million per year, increase the effective patent life and use 35 000 fewer laboratory animals.

would substantially reduce the time and costs of commercializing drugs. Patients, too, would gain by having faster access to improved medicines.

An even more comprehensive agreement on drug testing would allow a company to carry out only one set of scientific investigations, animal experiments and human trials. Afterwards it could apply to register a new drug anywhere in the world. Differences in clinical requirements would still survive because medical practice and social conditions vary so much in different parts of the world. Developing countries, though they have only limited research capacity of their own, would obtain access to a wider range of modern drugs and could be assured of greater competition between potential suppliers.

International agreement on standards of quality control and GMP is another area where progress is possible. Several industrialized countries have already taken modest steps in this direction but calls for harmonization will grow as the process of internationalization continues. Assistance to producers in developing countries could be a particularly effective way of helping this part of the industry to serve its customers better. Consumers in industrialized countries would also benefit since higher production standards would help to combat the growing problem of pirate drugs.

The prospects for policy harmonization in other areas are less promising. International agreements on licensing procedures are technically possible but would be extremely complex and time-consuming. Harmonization of drug pricing and reimbursement would not be possible except among countries where disease patterns and standards of living were roughly equivalent.

COMPETITION AND CORPORATE STRATEGIES

This book has argued that any broad shifts in competitive abilities will occur very gradually. Japanese producers have steadily improved their research capabilities while simultaneously shifting the focus from therapeutic fields with short product lives and slow-growing markets to more lucrative areas. Now that research parity has been achieved they will not want to pay the large sums which foreign partners charge to market their products outside Japan. An export drive is unlikely to materialize, however. Instead Japanese producers are building up a system of offshore research and distribution which complements their growing domestic strengths. That process will require another decade or more. Only then will the Japanese be able to compete on an equal footing with large integrated companies in Europe and the United States.

The pattern of foreign direct investment which exists today will not change significantly in the next ten years. Leading American and European companies have already established footholds in each other's territories and are moving into Japan. Meanwhile the Japanese are returning the compliment. Within the next decade every major pharmaceutical company will have a substantial presence in each of the world's major geographic markets and in many developing countries. The rising costs of R & D will require that these companies pursue international sales aggressively and that they maintain a world presence through all possible means.

Buried beneath these broad shifts in competitive performance are an infinite number of microeconomic decisions which are gradually changing the industry's landscape. The most important ones will be taken by the large integrated firms which are the primary source of new drugs. The profits these companies earn from innovation are falling as a result of competition from generics, patent erosion and other factors. Meanwhile the real costs of R & D continue to rise and the effects can only be partially offset by improvements in research techniques. Large integrated firms will therefore have to abandon several of the strategies that were common in the 1980s.

The wave of mergers and acquisitions that marked the end of the last decade has subsided but will be resumed during the 1990s. Integrated firms must be bigger if they are to compete on a global scale. Competition will also become more intense, not least because of the growing international presence of Japanese producers. So far, no single company has ever claimed more than 4 per cent of the world market. Nor has the list of the world's top companies changed greatly over the past several decades. These hallmarks of the industry could soon disappear. Some companies – those which can afford the exorbitant costs of research – will manage to build shares of 10 per cent or more. Others, operating on the theory that specialization will help in an industry becoming increasingly competitive, will be content to serve niche markets.

Several of the dominant companies will combine their research leadership with a greater interest in generic products. The existence of dual producers, coupled with greater government pressure for generic substitution, will lead the industry to adopt more aggressive tactics in its promotional campaigns. Such moves are inevitable but they will give rise to a wider debate on the ethics of marketing drugs. Individual firms are already being accused of conducting promotional drives which are disguised as educational or fact-finding campaigns. The post-market studies funded by various producers[6] and the close relationship which some have developed with national regulatory agencies will be sources of concern.

The industry's critics fear that a company's scientific judgement may become clouded as the scale of its promotional effort grows. Regulators will inevitably come under pressure to respond. They will have to create their own systems of post-market studies to evaluate the potential side-effects of drugs and to ensure that doctors are not heavily dependent on the industry's marketing men for information.[7]

More generally the industry's marketing strategies will become similar to those followed by producers of other consumer products, but the ethics of the industry cannot be ignored. A greater emphasis on marketing strategies will also lead to more extensive regulations in this field.

GLOBAL NEEDS AND MARKET PRIORITIES

Demographic trends, policy decisions and the microeconomics of the industry will be the main forces of change. Regrettably a fourth issue, the growing imbalance between demand and supply in developing countries, will probably be of less significance. The requirements of these countries are clearly growing at a pace which far outstrips their ability to produce or import the necessary medicines.

Conditions vary but the situation is gravest in parts of Africa and East Asia where basic health services are extremely limited. In these countries the mortality rates for infants and mothers are the world's highest, malnutrition is pervasive and life expectancy at birth is very low. Meanwhile, per capita consumption of pharmaceuticals has fallen to minimal levels – below $5 per year in many cases. These sums, although insufficient to meet the basic medicinal needs, are likely to decline further; spending on drugs may soon fall to below $1 per capita in a number of developing countries. Between 1.5 and 2.5 billion people are at present estimated to have little or no regular access to essential drugs (UNDP, 1991, p. 63) but that number will rise rapidly in the 1990s.

There are many reasons for such an alarming trend and only two of the more important are mentioned here. One is that many consumers are unable to obtain the necessary drugs at affordable prices. The other is that the types of drugs being brought to market do not closely match the medical needs in poor countries. The mismatch reflects a combination of factors: the income gap between rich and poor countries, the differential effects of demographic trends and adjustments in the industry's research priorities.

The first of these problems is part of a more general economic malaise. Substantial increases in purchasing power are contingent on a range of actions which go beyond the subject of this book. Certain aspects, however, are more directly linked with the manufacture and distribution of drugs. Though the evidence is limited, some analysts believe that substantial savings are possible if governments change their methods of purchase and distribution. One proposal is to reject some of the more modern, high-technology treatments in favour of simpler methods which are cheaper and equally effective.[8] Other analysts maintain that improved storage and distribution can yield large savings. For example, the wastage in immunization campaigns is around one-third and even a 20 per cent reduction would lead to a significant fall in costs. A third – and more sweeping – proposal calls for drastic changes in patterns of consumption. In explaining their reasoning, investigators also cite anecdotal evidence that some countries spend excessively on non-essential drugs. With careful regulation, essential drugs are estimated to cost around $1 per person and an even more basic list could be provided for $0.25 (UNDP, 1991). Such drugs would still be affordable even at the depressed levels of spending which are expected in the 1990s. The criticism of current procedures is probably accurate in selected cases. In others the pattern of consumption conforms rather closely with WHO's essential drug list (see Box 8.4). That outcome is achieved despite many difficulties, such as the brand-name system (see Chapter 7).

The second major issue raised here – the divergence between patterns of drug research and patterns of consumption in poor countries – has closer links with the pharmaceutical industry itself. Tropical dis-

Box 8.4   The pattern of drug consumption in Cameroon

WHO's list of essential drugs includes 275 medicinal chemicals which reflect the therapeutic needs of developing countries. Box table 8.4 relates the pattern of consumption in Cameroon to the WHO model list. Altogether the 20 best-selling product families accounted for 95 per cent of all sales on the private market. Twelve of these product families appear on the WHO list. Of the remaining eight, two are regarded by WHO as therapeutic alternatives, three are combinations containing active ingredients which appear on the WHO list and three are antibiotics used in the treatment of infections. The conclusion is that the country's therapeutic needs and consumption pattern coincide rather closely.

Box table 8.4   Consumption of essential drugs in Cameroon, 1987

Product family	WHO model list	Annual sales ($ thousand)	Market share
Ampicillin	[a]	1 577	11.9
Chloroquine	[a]	1 311	9.9
Acetylsalicylic Acid	[a]	1 075	8.1
Quinine	[a]	997	7.5
Mebendazole	[a]	816	6.1
Cotrimoxazole	[a]	671	5.0
Terpone[e]	[b]	623	4.7
Kanamycin	[b]	585	4.4
Gentamycin	[a]	583	4.4
Pyrantel pamoate	[a]	569	4.3
Troleandomycin	[b]	566	4.3
Tothema[e]	[c]	508	3.8
Bacampicillin	[d]	415	3.1
Doxycycline	[a]	366	2.8
Diclofenac sodium	[d]	354	2.7
Neo-Codion[e]	[c]	350	2.6
Spectinomycin	[a]	346	2.6
Sulfadoxamine	[a]	339	2.6
Hemostyl[e]	[c]	331	2.5
Albendazol	[a]	293	2.2
Others		613	4.6
Total		13 288	100.0

Notes
[a] Included in the WHO model list.
[b] Not included by WHO but used in the treatment of prevailing diseases.
[c] Major active ingredients are included in the WHO model list.
[d] Included as a therapeutic alternative.
[e] Registered trade names of different pharmaceutical companies.

Source: UNIDO.

eases, childhood diseases, tuberculosis and diarrhoea are some of the most common and dangerous afflictions in developing countries. The development of medications to treat sufferers has nevertheless lagged behind the industry's successes in other fields.

The methods of treatment for tropical diseases illustrate the situation. Most of the drugs used to combat these problems have been available for many years and were not developed through any coordinated research effort. Anthelmintics, for example, were obtained as a by-product of veterinary research while antimalarials were originally intended for military use. WHO's Programme for Research and Training in Tropical Diseases attempts to promote the search for NMEs to control diseases such as schistosomiasis, malaria, trypanosomiasis and leprosy. However the programme's budget is only about $30 million per year, a small sum in view of the fact that the research and marketing costs of launching an NME may be as high as $250 million.

More support for research into tropical diseases is obviously needed, although poor living conditions (rather than a shortage of effective medications) are the main source of tropical diseases. This fact means that the most effective form of assistance is greater amounts of aid to improve infrastructure, social services and so on. The industry itself can still play an important supporting role, and sometimes does (see Box 8.5).

In the case of other afflictions such as childhood diseases, tuberculosis and other infectious diseases, the pharmaceutical industry's potential contribution is much greater. More effective (and cheaper) treatments are clearly needed and are possible. Here the slow pace of progress is the result of market priorities. The industry's profile of

---

Box 8.5   Research on the diseases of developing countries

One example of the way the industry can respond to the needs of the developing countries is suggested by the recent agreement on production of Eflornithine. The drug, which was one of the NMEs approved by the FDA in 1990, is used for the treatment of African trypanosomiasis. It will not be marketed in the United States but will be manufactured there and in France and distributed at cost to African hospitals and health centres. Such an approach could serve as a model for other drugs which are badly needed by African countries.

future products is in the hands of research managers in large, integrated companies. This group is already making decisions about the drugs they hope to sell in the next decade. Table 8.3 shows the main types of drugs which were under development in 1989 and would be ready to market by the end of the century. Some – for example, anti-infectives – can find application among the younger populations in developing countries. Most, however, will be intended to treat chronic or degenerative diseases among the older (and richer) consumers in industrialized countries.

Such an orientation is not new. Commercial considerations explain the research bias. Drugs to treat the diseases of the rich are more profitable than cures for the ailments common in developing countries. The same applies to the manager's predilection for research into diseases which require treatment but may not be curable: these, too, are relatively profitable since medication continues over long periods of time. There is a danger that the profile of available drugs will become even more skewed as integrated firms struggle to maintain profitability in the face of more intense competition. If this should happen, any improvements in efficiency or reductions in the costs of drugs would be achieved at the expense of an even greater research specialization on the diseases of the industrialized countries.

Table 8.3 NMEs in preclinical development, 1989

Category	Number	Share in total
Anti-infective drugs	779	17.9
Biotechnology products	577	13.2
Anti-cancer products	563	12.9
Neurological drugs	450	10.3
Cardiovasculars	378	8.7
Anti-parasitic products	50	1.1
Others	1 559	35.9
Total	4 356	100.0

Source: Scrip, *Yearbook* (1990).

Pressures of a different sort threaten to reinforce the shift in research priorities which market forces have initiated. The sorts of demographic trends described in this chapter are the most important, but certain policy shifts in the industrialized countries could also have an impact. As the number of people afflicted by addictive disorders associated with the use of cocaine, heroin and other narcotics grows, government officials in some industrialized countries will want new types of pharmaceuticals for treatment. Until now the potential medications for addictive disorders have been regarded as unprofitable by the industry. That will change as governments attempt to influence research priorities by offering more funding and policy support for such work.

In conclusion the prospects for any industry ultimately depend on the confidence of its consumers. Producers of pharmaceuticals have a special problem in view of the ethical and political issues which have always played a central role in the case of pharmaceuticals and their significance is accentuated by society's growing ambivalence towards the industry. Drug companies have not dealt with these concerns in an effective manner in the past. This task will only become more difficult if, in the future, the industry is seen to neglect the growing needs of the developing countries as well.

NOTES

1. This is not necessarily true for the elderly in developing countries where consumption patterns differ from those in industrialized countries. The former group have far less income to spend on medicines and are generally healthier. As a result the frequency of drug consumption among the elderly is substantially lower in developing countries than in industrialized ones.
2. Space does not permit a discussion of the latest trends in individual countries, though a few are significant. For example, growth in the German market has slowed in 1990–1 as a result of new pricing policies introduced in 1989. The pharmaceutical industry in the USSR is performing poorly owing to the disruption in traditional trading links and a deterioration in the general economic situation. In the developing countries, the liberalization of drug prices in Argentina and Brazil has improved growth prospects. Meanwhile the value of pharmaceutical sales in China has fallen by a quarter and the reasons are unclear.
3. Examples of policies being employed in developing countries include tighter controls on labelling, the compulsory use of essential drugs (see Box 2.5) and

continued efforts to regulate promotional campaigns and weaken the brand-name system (see Chapter 6). The measures being adopted in industrialized countries include reductions in reimbursement levels, ceilings on reimbursable prices and a variety of other incentives to encourage doctors, pharmacists and patients to purchase generic products. For further discussion, see Chapters 2 and 6.

4. Some policies that are currently enforced discourage generic competition and may eventually have to be replaced. One example is the use of fixed-profit margins for pharmacists. When these are specified as a percentage of the selling price, the pharmacist has an incentive to dispense the most expensive products. Other regulations have similar, though unintended, effects. Some countries require that a product which is frequently prescribed must be stocked and cannot be replaced by another. As a result, pharmacists will choose to dispense an established product even if a generic is prescribed.

5. An example is provided by anti-hypertensives, beta-blockers, calcium antagonists and ACE-inhibitors. These drugs are distinct but they are all used in the treatment of high blood pressure.

6. The idea behind post-market studies is to identify any adverse effects of a drug in a normal population, rather than the one carefully selected for clinical trials. There is a need for such studies but regulators have yet to develop a system to handle any adverse reports which emerge in the post-marketing period. This role is being increasingly assumed by the firms which develop the new products. Because they pay doctors to conduct the study, critics regard the practice as a covert form of promotion.

7. If doctors are too dependent on the supplier's sales force for information, no debate on side-effects may take place. This could be why WHO has gone so far as to propose that scientific sales promotion should be financed by the state, as practised in the (former) centrally planned economies.

8. An example is found in the treatment of diarrhoeal diseases. Intravenous rehydration therapy is the modern method of treatment but oral rehydration is just as effective and its use could cut costs by about 90 per cent.

Statistical Appendix

Global figures on production and consumption were derived from national data for more than 140 countries/areas (for the country coverage, see Table A.1). Unless otherwise mentioned, 'world' refers to these countries/areas.[1] The data refer to pharmaceutical preparations or medicaments. These drugs may be supplied either in bulk or in retail packs for human and veterinary use.

Whenever possible the global and regional estimates are expressed in real terms. When data on real consumption (or total domestic sales) were not available, figures were estimated on the basis of apparent consumption, which is defined as gross output plus imports less exports, and ignores annual changes in stock. Similarly, if data on real production were not available, production was approximated by real consumption plus net exports.

Data on real consumption and production are generally reported at manufacturers' selling prices. Data on apparent consumption, however, were derived from statistics with mixed validation (for example, producers' prices, factor prices, cost, insurance, freight (cif) prices or free on board (fob) prices). For purposes of international comparisons, these estimates were converted to United States dollars by applying period averages of market exchange rates.

Any measurements expressed in current value terms can be distorted as a result of high rates of inflation and large fluctuations in official exchange rates. Furthermore exchange rates of some countries have been fixed for several years, while those of others have fluctuated (some freely, some under strict government control). In order to avoid these distortions and to enhance the international comparability of data, all national statistics on production and consumption were deflated using a manufacturing value added (MVA) deflator, GDP deflator, the consumer price index or an output deflator for the whole chemical industry. After deflation, values were converted to constant 1980 dollars using 1980 average period rates of exchange for the respective country.

The unavailability of production and price indices for pharmaceuticals meant that real values of production and consumption for many of the countries had to be approximated with the help of a deflator which was relatively broad in scope. Use of such deflators poses problems since, in many countries, the prices of pharmaceuticals are fixed or strictly controlled by the governments. Therefore price changes may not necessarily follow a more broadly defined pattern of inflation.

In addition to exchange rate problems, international comparisons of the pharmaceutical industry are subject to various other statistical difficulties which arise from national differences in scope of the industry, product classification, statistical definitions and coverage, prices and so forth. Although every effort has been made to increase cross-country comparability, empirical results should be regarded as approximations.

NOTE

1. East European countries and the USSR do not appear in Table A.1. For these countries, cross-country aggregates and averages are shown whenever possible. The country group includes Albania, Bulgaria, Czechoslovakia, German Democratic Republic, Hungary, Poland, Romania and the USSR.

Table A.1 Selected indicators for production and consumption of pharmaceutical preparations, by countries, selected years

Country/area	Share in world total[a] (Percentage)				Consumption per capita (Constant 1980 US$)		Consumption in GDP (Percentage)		Percentage ratio of					
	production		consumption						production to consumption		imports to consumption		exports to production[b]	
	1975	1990	1975	1990	1975	1990	1975	1990	1975	1989	1975	1989	1975	1989
Developed countries														
North America														
Canada	1.15	1.34	1.23	1.49	37.1	83.4	0.4	0.6	95.9	91.0	8.0	12.0	4.1	3.3
United States	19.24	21.31	19.22	21.55	60.8	128.2	0.5	0.9	102.8	100.2	0.4	2.0	3.1	2.2
EC														
Belgium	1.04	0.94	1.11	0.82	77.2	121.7	0.7	0.7	96.4	116.4	54.7	59.9	53.0	65.6
Denmark	0.42	0.45	0.30	0.22	40.6	64.1	0.3	0.3	141.9	206.4	42.8	61.8	59.7	81.5
France	6.53	6.34	5.64	5.69	73.2	150.2	0.7	0.9	118.7	112.8	1.0	10.4	16.6	20.6
Germany, Fed. Rep.	7.79	6.50	6.98	6.08	77.2	148.9	0.7	0.8	114.6	108.4	8.9	12.5	20.5	19.3
Greece	0.50	0.29	0.63	0.35	47.7	51.7	1.3	1.0	80.5	85.3	22.7	25.6	4.0	12.8
Ireland	0.02	0.26	0.08	0.11	17.8	45.0	0.4	0.6	31.2	231.7	133.6	112.0	207.8	105.2
Italy	3.91	3.69	3.96	3.92	48.9	101.5	0.8	0.9	101.2	95.5	7.2	11.3	8.3	7.1
Luxembourg	–	–	0.03	0.02	65.2	99.4	0.6	0.5	–	–	100.0	100.0	...	...
Netherlands	0.68	0.64	0.72	0.70	35.9	70.7	0.3	0.4	96.7	92.7	63.8	74.2	62.6	72.2
Portugal	0.31	0.29	0.39	0.34	29.2	48.9	1.3	1.3	80.7	87.9	23.8	16.5	5.7	4.9
Spain	3.16	1.72	3.22	1.76	61.9	66.4	1.2	0.8	100.7	98.7	0.9	7.4	1.6	6.2
United Kingdom	4.24	3.14	2.97	2.52	36.0	65.5	0.4	0.4	146.5	126.3	10.6	26.7	39.0	42.0

Other Europe														
Austria	0.29	0.26	0.47	0.41	42.5	81.9	0.5	0.5	62.8	64.4	55.5	56.2	29.2	32.0
Finland	0.16	0.16	0.31	0.24	45.0	70.6	0.5	0.4	54.2	67.1	48.1	48.0	4.3	22.5
Iceland	–	–	0.01	0.01	37.8	75.5	0.4	0.3	–	–	100.0	100.0	...	...
Norway	0.07	0.07	0.21	0.17	35.1	60.0	0.3	0.3	33.8	43.5	71.0	66.4	14.3	22.8
Sweden	0.48	0.75	0.62	0.51	51.5	90.8	0.4	0.4	80.5	149.8	57.7	55.2	47.4	70.1
Switzerland	1.68	1.34	0.73	0.49	77.5	111.9	0.5	0.4	237.9	276.9	36.1	51.3	73.1	82.4
Japan	14.19	22.30	15.02	23.03	92.0	276.6	1.1	1.6	97.0	98.1	3.6	2.1	0.6	0.3
Others														
Australia	0.68	0.51	0.80	0.67	40.4	58.9	0.4	0.4	86.8	78.3	20.8	29.3	8.7	9.8
Israel	0.13	0.08	0.16	0.11	31.8	35.5	0.6	0.4	81.6	74.9	23.6	43.3	6.4	24.3
New Zealand	0.02	0.05	0.12	0.13	27.5	55.7	0.4	0.5	18.1	43.1	84.1	60.6	12.4	8.6
South Africa	0.45	0.44	0.54	0.54	14.5	22.7	0.6	0.8	85.6	83.7	18.6	17.3	4.9	1.2
Developing countries														
Latin America and Caribbean														
Antigua-Barbuda	–	–	–	–	8.6	13.5	0.8	0.4	–	–	100.0	100.0	...	...
Argentina	1.50	1.68	1.55	1.74	40.6	79.7	1.4	0.7	99.2	98.1	1.8	2.9	1.0	1.1
Bahamas	–	–	–	–	12.5	31.2	0.3	0.4	–	–	100.0	100.0	...	...
Barbados	–	–	0.01	0.01	26.2	41.0	1.0	1.0	0.2	22.2	108.5	80.0	4 350.0	10.0
Belize	–	–	–	–	8.8	11.3	0.8	0.9	71.9	58.8	30.2	43.2	2.9	3.3
Bermuda	–	–	0.01	0.01	156.5	162.7	...	...	–	–	100.0	100.0	...	...

Country/area	Share in world total[a] (Percentage)				Consumption per capita (Constant 1980 US$)		Consumption in GDP (Percentage)		Percentage ratio of					
	production		consumption						production to consumption		imports to consumption		exports to production[b]	
	1975	1990	1975	1990	1975	1990	1975	1990	1975	1989	1975	1989	1975	1989
Bolivia	0.02	0.01	0.05	0.02	7.5	4.0	0.9	0.6	38.7	57.7	61.3	42.3	–	–
Brazil	1.91	1.05	1.97	1.07	12.5	10.5	0.9	0.7	99.7	99.8	0.6	1.8	0.3	1.6
Br. Virgin Islands	–	–	–	–	4.2	10.2	0.2	0.2	–	–	100.0	100.0	...	...
Chile	0.21	0.14	0.23	0.18	14.9	20.4	1.2	0.7	95.8	80.6	4.5	20.3	0.3	1.1
Colombia	0.41	0.41	0.42	0.42	12.5	19.6	1.1	1.1	99.9	97.8	5.3	4.4	5.2	2.3
Costa Rica	0.05	0.04	0.06	0.05	22.2	24.6	1.5	1.4	72.9	81.3	60.4	62.5	45.7	53.8
Dominica	–	–	–	–	10.7	12.6	1.7	1.0	–	–	100.0	100.0	...	...
Dominican Rep.	0.01	0.03	0.06	0.05	7.6	10.9	0.7	2.0	18.1	51.1	82.4	50.0	2.6	2.1
Ecuador	0.09	0.03	0.16	0.10	15.7	13.5	1.5	1.2	58.3	34.9	43.9	66.4	3.7	3.8
El Salvador	0.05	0.01	0.09	0.03	15.2	8.0	1.9	1.5	59.1	54.9	59.3	62.2	31.3	31.1
French Guiana	–	–	–	–	5.9	50.2	...	1.1	–	–	100.0	100.0	...	...
Grenada	–	–	–	–	8.1	12.6	1.1	2.2	–	–	100.0	100.0	...	...
Guadeloupe	–	–	0.01	0.02	27.6	79.3	0.8	1.1	–	–	100.0	100.0	...	...
Guatemala	0.09	0.07	0.10	0.07	11.3	11.6	1.1	1.1	96.3	105.1	48.1	41.2	46.1	44.1
Guyana	0.01	–	0.01	–	10.7	6.6	1.0	1.1	49.8	80.0	74.1	40.0	48.0	25.0
Haiti	–	–	0.01	0.01	1.4	2.0	0.7	0.7	19.4	61.8	82.7	40.8	10.6	4.1
Honduras	0.02	0.01	0.05	0.01	11.2	4.0	1.9	0.7	33.9	58.6	66.1	42.1	0.1	1.2
Jamaica	–	–	0.02	0.02	6.1	10.0	0.3	0.7	11.1	27.9	92.9	80.1	35.6	28.7
Martinique	–	–	0.02	0.02	36.8	105.4	0.9	2.2	–	–	100.0	100.0	...	...
Mexico	1.52	1.09	1.54	1.11	17.0	18.6	0.8	0.7	101.4	98.8	0.4	2.3	1.8	1.0
Montserrat	–	–	–	–	25.6	27.6	1.7	1.3	–	–	100.0	100.0	...	...

Neth. Antilles	–	–	0.01	0.01	23.4	50.4	0.4	0.5	18.1	28.6	100.0	100.0	..	..
Nicaragua	0.01	0.02	0.07	0.08	19.1	30.0	..	..	20.8	37.0	81.9	71.4	–	–
Panama	0.01	0.01	0.04	0.03	16.2	19.7	1.1	0.8	84.4	88.1	79.3	73.1	0.5	27.3
Paraguay	0.05	0.05	0.06	0.05	14.2	18.8	1.3	0.4	96.2	92.8	15.6	12.1	–	0.3
Peru	0.26	0.22	0.27	0.25	12.3	16.3	1.1	0.5	..	..	4.4	7.7	0.7	0.5
Puerto Rico	3.19	2.65	0.13	0.21	30.3	85.0	0.8	1.4	..	..	..	..	..	..
St Kitts–Nevis	–	–	–	–	9.5	17.8	1.1	1.0	–	–	100.0	100.0	..	..
St Lucia	–	–	–	–	5.5	7.9	0.8	0.7	–	–	100.0	100.0	..	..
St Vincent	–	–	–	–	0.7	3.4	0.1	0.4	–	–	100.0	100.0	..	..
Suriname	–	–	–	–	10.2	6.0	0.4	1.1	–	–	100.0	100.0	..	..
Trinidad–Tobago	0.01	–	0.03	0.02	17.9	18.2	0.4	0.9	23.5	22.9	78.5	84.3	8.7	31.5
US Virgin Islands	–	–	–	–	1.2	1.3	..	..	–	–	100.0	100.0	..	..
Uruguay	0.04	0.07	0.05	0.08	12.7	40.0	0.4	1.2	77.4	85.0	23.0	17.2	0.6	2.7
Venezuela	0.59	0.29	0.63	0.31	34.2	23.2	0.8	0.4	95.6	93.5	4.5	6.7	0.2	0.2
North Africa														
Algeria	0.05	0.06	0.35	0.29	14.8	16.8	0.8	1.2	15.3	19.7	84.7	80.4	0.1	0.4
Egypt	0.14	0.06	0.17	0.13	3.2	3.7	0.9	0.3	86.3	43.9	16.2	64.1	2.9	18.2
Libyan Arab	–	–	0.06	0.06	16.8	19.3	0.2	0.5	–	–	100.0	100.0	..	..
Morocco	0.13	0.17	0.17	0.19	6.8	11.2	0.8	0.8	76.8	90.3	24.4	13.7	1.7	4.4
Sudan	0.15	0.06	0.21	0.09	9.1	5.0	2.1	..	71.3	70.4	28.7	29.6	–	0.1
Tunisia	–	0.03	0.06	0.11	7.0	20.0	0.6	1.3	8.6	30.3	93.2	69.9	20.8	0.9
Other Africa														
Angola	0.01	0.03	0.03	0.03	3.4	5.0	0.7	0.7	43.4	84.6	56.8	15.4	0.5	–
Benin	–	–	0.01	0.01	2.4	2.6	0.8	0.7	–	29.1	100.0	70.9	..	–
Botswana	–	–	–	0.01	4.5	13.5	0.7	0.7	–	–	100.0	100.0	..	..
Burkina-Faso	–	–	0.01	0.01	1.2	2.0	0.7	0.7	–	–	100.0	100.0	..	..
Burundi	–	–	–	–	0.7	0.9	0.4	0.5	–	–	100.0	100.0	..	..

Country/area	Share in world total[a] (Percentage)				Consumption per capita (Constant 1980 US$)		Consumption in GDP (Percentage)		Percentage ratio of					
	production		consumption						production to consumption		imports to consumption		exports to production[b]	
	1975	1990	1975	1990	1975	1990	1975	1990	1975	1989	1975	1989	1975	1989
Cameroon	0.02	0.01	0.05	0.06	4.5	7.8	0.7	1.0	42.9	10.0	57.1	90.4	–	4.0
Cape Verde	–	–	–	–	2.1	5.2	0.7	0.7	28.2	57.3	71.8	43.2	–	0.8
C. African Rep.	–	–	–	–	1.5	2.5	0.4	0.6	–	–	100.0	100.0	...	...
Chad	–	–	–	–	0.7	1.1	0.3	0.4	–	–	100.0	100.0	...	...
Comoros	–	–	–	–	1.4	3.7	0.5	0.6	–	–	100.0	100.0	...	...
Congo	–	–	0.01	0.02	4.7	13.1	0.5	1.5	–	–	100.0	100.0	...	...
Côte d'Ivoire	0.03	0.01	0.09	0.08	9.0	9.5	0.7	1.3	30.5	14.6	69.6	86.3	0.3	6.3
Equatorial Guinea	–	–	–	–	0.5	0.4	...	...	–	–	100.0	100.0	...	–
Ethiopia	0.02	0.01	0.05	0.02	1.0	0.7	0.6	0.9	43.8	38.8	56.2	61.2	–	–
Gabon	–	–	0.01	0.01	8.8	15.2	0.1	0.6	–	–	100.0	100.0	...	...
Gambia	–	–	–	–	3.4	2.6	0.9	0.5	–	–	100.0	100.0	...	...
Ghana	0.05	0.06	0.10	0.07	6.8	6.7	0.8	1.3	48.8	83.2	52.2	17.3	2.1	0.6
Guinea-Bissau	–	–	–	–	0.5	1.4	0.2	1.2	–	–	100.0	100.0	...	...
Kenya	0.03	0.03	0.07	0.05	3.4	3.2	0.8	0.5	52.1	58.0	52.5	42.5	8.8	0.9
Lesotho	...	...	–	–	1.5	2.1	0.7	0.7	...	...	...	...	...	...
Liberia	–	–	0.01	–	3.9	2.4	0.6	0.7	12.3	27.3	87.9	72.8	1.6	0.4
Madagascar	–	0.01	0.02	0.03	2.2	3.2	0.6	1.8	–	52.7	100.0	47.4	...	0.2
Malawi	0.01	0.01	0.02	0.01	2.2	2.1	1.1	2.0	74.8	83.4	30.9	21.6	7.7	6.0
Mali	–	–	0.01	0.01	1.5	1.7	0.7	0.7	–	30.2	100.0	69.8	–	–
Mauritania	–	–	–	0.01	2.4	4.3	0.4	1.4	–	–	100.0	100.0	...	...
Mauritius	–	–	0.01	0.01	6.5	10.4	0.5	0.5	–	8.7	100.0	91.3	...	–

Mozambique	—	—	0.10	0.01	6.2	1.5	0.4	1.5	—	—	100.0	100.0	...	...
Niger	0.02	0.01	0.02	0.02	3.1	3.8	0.7	0.7	85.9	54.8	14.1	45.2	—	—
Nigeria	0.38	0.13	0.84	0.20	8.5	2.7	0.6	0.2	46.0	62.4	54.0	37.6	—	0.1
Reunion	—	—	0.03	0.04	48.3	88.7	1.4	2.2	—	—	100.0	100.0	...	...
Rwanda	—	—	—	—	0.4	0.5	0.2	0.2	—	—	100.0	100.0	...	...
Senegal	0.01	0.01	0.03	0.02	3.9	5.1	0.7	0.7	41.6	24.5	62.0	93.8	8.8	74.9
Seychelles	—	—	—	—	10.0	24.7	0.7	0.6	76.6	69.9	23.4	30.1	—	—
Sierra Leone	0.01	0.01	0.02	0.01	4.4	2.4	1.3	1.9	61.4	90.0	38.6	9.9	0.1	—
Somalia	0.01	—	0.02	0.01	3.1	2.5	0.7	0.7	45.3	43.1	54.7	56.8	—	—
Swaziland	—	—	—	—	7.4	9.5	0.7	1.0	—	—	100.0	100.0	...	...
Togo	—	—	0.01	0.01	3.0	4.5	0.6	0.8	—	—	100.0	100.0	...	...
Uganda	—	...	0.04	0.03	2.7	2.7	...	...	—	...	100.0	...	...	...
U.R. of Tanzania	0.03	0.03	0.08	0.05	3.7	2.7	1.3	1.6	31.9	60.9	68.1	39.1	—	1.3
Zaire	0.11	0.04	0.19	0.08	5.8	3.3	0.7	0.7	60.3	45.3	39.7	55.3	—	—
Zambia	0.01	0.01	0.04	0.02	5.8	3.0	0.7	0.2	29.4	33.3	70.6	66.7	—	—
Zimbabwe	0.03	0.02	0.04	0.03	4.2	4.3	0.5	0.5	97.0	84.3	6.5	25.7	3.5	11.9
South & East Asia and Oceania														
Afghanistan	0.01	—	0.03	0.01	1.2	0.8	...	...	42.7	54.6	57.3	45.5	—	—
Bangladesh	0.10	0.07	0.11	0.08	1.0	1.1	0.6	0.7	89.9	83.8	10.4	16.3	0.3	0.2
Bhutan	—	—	—	—	0.6	1.6	0.6	0.7	—	—	100.0	100.0	...	...
Brunei	0.01	—	0.01	0.01	67.2	36.7	0.6	0.6	85.9	65.7	14.1	34.3	—	0.1
Fiji	—	—	—	0.01	5.3	12.6	0.3	1.1	—	60.2	100.0	39.9	...	0.3
Fr. Polynesia	—	—	—	0.01	25.8	48.9	0.4	0.7	—	—	100.0	100.0	...	...
Hong Kong	0.04	0.04	0.16	0.18	24.2	45.6	0.8	0.4	23.6	22.2	89.5	87.3	55.3	42.7
India	0.93	1.29	0.98	1.25	1.1	2.2	0.6	0.5	96.9	104.3	6.5	5.4	3.5	9.4
Indonesia	0.36	0.46	0.39	0.47	2.0	3.9	0.6	0.5	94.6	98.8	5.8	2.0	0.4	0.9
Iran (Islamic Rep.)	0.25	0.81	0.82	0.96	16.7	25.0	0.5	0.3	31.3	85.3	69.0	14.8	1.0	0.2

Country/area	Share in world total[a] production (Percentage)		consumption (Percentage)		Consumption per capita (Constant 1980 US$)		Consumption in GDP (Percentage)		Percentage ratio of production to consumption		imports to consumption		exports to production[b]	
	1975	1990	1975	1990	1975	1990	1975	1990	1975	1989	1975	1989	1975	1989
Macau	–	–	0.01	0.01	15.6	21.2	...	...	40.5	29.4	62.8	98.4	8.3	94.7
Malaysia	0.02	0.04	0.07	0.09	4.0	7.8	0.4	0.3	27.5	49.5	87.5	62.5	54.4	24.2
Maldives	–	–	–	–	0.7	2.0	0.2	0.5	–	–	100.0	100.0	...	...
Myanmar	0.02	0.01	0.03	0.01	0.6	0.3	...	...	77.0	60.5	23.0	39.7	–	0.4
Nepal	–	–	0.01	0.01	0.5	1.1	0.3	0.6	–	–	100.0	100.0	...	...
New Caledonia	–	–	–	0.01	22.9	84.2	0.3	1.8	–	–	100.0	100.0	...	...
Pakistan	0.14	0.33	0.18	0.42	1.7	5.1	0.7	1.1	80.9	79.1	20.5	21.6	1.8	0.9
Papua N. Guinea	0.01	0.01	0.02	0.02	5.7	6.3	0.7	0.7	70.4	79.5	29.6	20.5	–	–
Philippines	0.39	0.29	0.41	0.32	6.6	7.7	1.0	1.1	98.5	89.7	1.6	11.3	–	1.2
Rep. of Korea	0.87	0.97	0.89	1.00	17.3	33.9	1.6	0.9	100.3	98.3	1.2	2.5	1.5	0.9
Samoa	–	–	–	–	0.8	2.3	0.1	0.3	–	–	100.0	100.0	...	...
Singapore	0.01	0.03	0.02	0.04	7.6	20.1	0.2	0.2	58.2	83.5	314.3	205.0	468.4	225.8
Solomon Islands	–	–	–	–	2.6	5.7	0.7	0.7	72.3	82.4	27.9	17.6	0.4	0.1
Sri Lanka	–	–	0.01	0.02	0.6	1.8	0.3	0.5	25.8	16.7	74.4	84.1	0.8	4.5
Taiwan Province	0.22	0.34	0.27	0.39	11.3	28.3	...	...	86.8	89.4	16.7	12.6	4.0	2.3
Thailand	0.24	0.21	0.34	0.25	5.6	6.6	1.0	0.5	71.7	87.6	29.0	14.3	1.0	2.1
Tonga	–	–	–	–	2.3	2.1	0.4	0.3	–	–	100.0	100.0	...	...
Vanuatu	–	–	–	–	5.2	6.6	0.3	0.5	–	–	100.0	100.0	...	...
China	5.57	3.55	5.70	3.58	4.3	4.8	1.7	0.6	100.4	100.5	0.7	3.7	1.1	4.2

Others

Bahrain	–	–	0.01	–	24.5	15.5	0.3	0.3	–	–	100.0	100.0	...	91.6
Cyprus	0.01	0.01	0.02	0.02	17.9	46.3	1.0	0.7	39.9	49.2	60.6	95.9	1.3	4.5
Dem. Yemen	–	–	–	–	1.7	3.3	0.7	1.0	16.4	9.9	83.7	90.5	0.6	–
Iraq	0.16	0.15	0.22	0.28	13.7	21.6	0.7	0.6	74.9	54.9	25.2	45.2	0.1	92.6
Jordan	–	0.05	0.02	0.04	5.0	13.5	0.7	1.1	2.1	122.7	153.9	90.9	2 721.5	94.3
Kuwait	–	0.01	0.06	0.03	39.6	24.9	0.2	0.3	–	25.0	100.0	98.6	...	...
Lebanon	–	–	0.03	0.03	8.2	15.4	0.7	1.1	–	–	100.0	100.0	...	...
Malta	–	–	0.01	0.01	17.3	35.4	0.8	0.6	–	14.8	100.0	129.0	...	296.1
Oman	–	–	0.01	0.01	10.9	13.7	0.2	0.3	–	–	100.0	100.0	...	...
Qatar	–	–	0.01	–	31.8	20.1	0.1	0.2	–	–	100.0	100.0	...	...
Saudi Arabia	–	0.03	0.15	0.45	13.9	47.0	0.2	0.8	–	7.3	100.0	92.9	...	2.9
Syrian Arab Rep.	0.01	0.04	0.16	0.05	14.4	5.6	1.2	1.9	7.2	84.1	92.8	15.9	0.7	–
Turkey	0.53	0.44	0.54	0.53	9.3	14.1	0.8	1.2	100.0	83.6	0.5	19.9	0.5	4.2
U. Arab Emirates	0.07	0.04	0.10	0.07	138.2	68.8	0.5	0.5	75.0	55.5	25.4	48.7	0.5	7.5
Yemen	–	0.01	0.06	0.04	8.2	7.4	2.4	0.7	–	21.3	100.0	78.7	...	–
Yugoslavia	1.35	0.54	1.35	0.24	43.1	15.2	1.9	1.1	103.1	222.6	4.7	16.1	7.6	62.3

Notes

a World totals include Eastern European countries and USSR, and shares were calculated at constant 1980 prices.

b For some countries/areas, the percentage ratios may exceed 100 per cent owing to possible inclusion of re-exports.

Source: Based on data reported by national pharmaceutical manufacturers' associations; IMS, *World Drug Market Manual*, various issues; United Nations trade tapes; estimates made by UNIDO.

Table A.2   *Relative growth and importance of the pharmaceutical industry in terms of value added in constant prices, by country, 1975 and 1988*

Country/area	Relative growth index for the period 1975–88		Percentage share of the pharmaceutical industry in total chemical production		Index of relative specialization within total manufacturing	
	Relative to the growth of total chemical production	Relative to the growth of MVA	1975	1988	1975	1988
Developed countries						
North America						
Canada	1.34	2.36	17.2	21.0	0.68	0.93
United States	2.92	1.34	17.5	25.6	1.15	1.14
EC						
Belgium	1.36	1.95	11.9	13.7	0.57	0.63
Denmark	1.76	3.52	24.8	35.2	1.10	1.69
Germany, Fed. Rep.	1.08	3.29	15.4	16.2	0.71	0.95
Greece	0.99	0.34	19.5	19.5	0.79	0.59
Luxembourg	...	...	–	–	–	–
Netherlands	...	...	...	13.1	...	0.75
Portugal	0.79	1.22	21.5	19.3	0.55	0.51
Spain	1.19	1.14	19.7	20.5	0.72	0.63
United Kingdom	2.86	15.91	15.3	23.4	0.92	1.46
Other Europe						
Austria	5.52	2.10	10.9	18.2	0.33	0.38
Finland	2.06	1.14	8.8	11.4	0.33	0.30
Iceland	...	...	–	–	–	–
Norway	4.14	23.79	4.5	12.5	0.14	0.44
Sweden	1.87	6.87	17.1	24.4	0.60	0.99
Japan	1.38	0.93	24.5	30.0	1.29	1.04
Others						
Australia	0.89	1.24	17.6	17.0	0.59	0.53
Israel	3.73	7.35	10.3	20.0	0.44	0.83
New Zealand	...	3.26	10.3	...	0.23	0.28

| Country/area | Relative growth index for the period 1975–88 | | Percentage share of the pharmaceutical industry in total chemical production | | Index of relative specialization within total manufacturing | |
	Relative to the growth of total chemical production	Relative to the growth of MVA	1975	1988	1975	1988
Developing countries						
Latin America and Carribean						
Antigua–Barbuda	...	...	–	–	–	–
Argentina	0.89	...	28.7	26.4	1.22	2.28
Barbados	...	...	–	–	–	–
Belize	...	...	–	–	–	–
Bermuda	...	...	–	–	–	–
Bolivia	...	...	...	25.9	...	0.60
Brazil	...	...	26.7	13.1	1.33	0.72
Brit. Virgin Isl.	...	...	–	–	–	–
Chile	4.96	0.46	25.2	29.4	1.75	1.19
Colombia	0.93	0.94	27.8	27.1	1.42	1.18
Costa Rica	1.70	1.26	20.5	25.5	0.87	0.83
Dominica	...	...	–	–	–	–
Dominican Rep.	...	...	5.1	...	0.14	...
Ecuador	0.12	0.15	38.5	23.0	0.78	0.44
El Salvador	...	...	23.9	...	1.77	2.25
French Guiana	...	...	–	–	–	–
Grenada	...	...	–	–	–	–
Guadeloupe	...	...	–	–	–	–
Guatemala	...	0.43	37.3	46.8	0.78	0.57
Honduras	0.36	0.78	35.2	20.9	0.60	0.47
Martinique	...	...	–	–	–	–
Mexico	0.42	0.90	23.7	15.2	1.25	1.03
Montserrat	...	...	–	–	–	–
Neth. Antilles	...	...	–	–	–	–
Nicaragua	...	...	...	...	...	0.54
Panama	1.06	3.12	14.9	15.6	0.54	0.78
Peru	...	...	23.1	20.8	1.13	0.71
St Lucia	...	...	–	–	–	–
St Vincent	...	...	–	–	–	–
Suriname	...	...	–	–	–	–
Trinidad–Tobago	...	...	...	73.1	...	0.82

| Country/area | Relative growth index for the period 1975–88 | | Percentage share of the pharmaceutical industry in total chemical production | | Index of relative specialization within total manufacturing | |
	Relative to the growth of total chemical production	Relative to the growth of MVA	1975	1988	1975	1988
Turks & Caicos	...	...	–	–	–	–
US Virgin Isl.	...	...	–	–	–	–
Uruguay	...	...	...	38.7	...	1.70
Venezuela	...	...	23.0	5.5	1.15	0.27
North Africa						
Egypt	...	...	41.0	9.1	1.20	0.42
Libyan Arab	...	...	–	–	–	–
Tunisia	...	...	13.3	...	0.81	...
Other Africa						
Botswana	...	...	–	–	–	–
Burkina-Faso	...	...	–	–	–	–
Cape Verde	...	...	–	–	–	–
C. African Rep.	...	...	–	–	–	–
Chad	...	...	–	–	–	–
Comoros	...	...	–	–	–	–
Congo	...	...	–	–	–	–
Djibouti	...	...	–	–	–	–
Equatorial Guin.	...	...	–	–	–	–
Gabon	...	...	–	–	–	–
Gambia	...	...	–	–	–	–
Ghana	...	...	24.5	...	0.61	...
Guinea-Bissau	...	...	–	–	–	–
Kenya	1.04	0.42	14.6	14.8	0.90	0.47
Madagascar	...	...	–	28.2	–	0.47
Malawi	...	...	74.2	–	1.82	–
Mauritania	...	...	–	–	–	–
Mauritius	...	...	–	3.1	–	0.06
Nigeria	...	...	19.0	...	1.63	...
Reunion	...	...	–	–	–	–
Rwanda	...	...	–	–	–	–
Sao Tome & Prin.	...	...	–	–	–	–
Senegal	...	...	–	–	–	–
Seychelles	...	...	–	–	–	–

Country/area	Relative growth index for the period 1975–88		Percentage share of the pharmaceutical industry in total chemical production		Index of relative specialization within total manufacturing	
	Relative to the growth of total chemical production	Relative to the growth of MVA	1975	1988	1975	1988
Sierra Leone	...	...	–	–	–	–
Togo	...	...	–	–	–	–
U.R. of Tanzania	...	...	–	1.1	–	0.03
Uganda	...	...	–	–	–	–
Zambia	...	24.06	2.1	14.2	0.10	0.44
South and East Asia and Oceania						
Bangladesh	0.89	2.70	38.7	33.9	0.80	1.43
Brunei	...	...	–	–	–	–
Fiji	...	...	–	–	–	–
Fr. Polynesia	...	...	–	–	–	–
Hong Kong	0.85	0.67	21.2	18.9	0.20	0.12
India	0.58	0.42	23.7	18.7	0.93	0.51
Indonesia	1.18	1.17	21.1	26.2	0.49	0.51
Iran (Islamic Rep.)	...	...	24.1	...	0.86	...
Malaysia	0.65	1.22	6.1	3.4	0.14	0.14
Maldives	...	...	–	–	–	–
Nepal	...	...	...	...	...	0.43
New Caledonia	...	...	–	–	–	–
Papua N. Guinea	...	...	–	–	–	–
Philippines	3.84	2.14	23.5	36.9	0.66	0.78
Rep. of Korea	0.93	0.75	27.7	25.6	1.67	1.03
Samoa	...	...	–	–	–	–
Singapore	...	...	44.5	...	1.09	...
Solomon Islands	...	...	–	–	–	–
Sri Lanka	...	...	...	16.1	...	0.16
Thailand	...	...	40.7	8.5	1.20	0.33
Tonga	...	...	–	–	–	–
Vanuatu	...	...	–	–	–	–
Others						
Bahrain	...	...	–	–	–	–
Cyprus	...	...	–	12.0	–	0.18
Dem. Yemen	...	...	–	–	–	–

Country/area	Relative growth index for the period 1975–88		Percentage share of the pharmaceutical industry in total chemical production		Index of relative specialization within total manufacturing	
	Relative to the growth of total chemical production	*Relative to the growth of MVA*	*1975*	*1988*	*1975*	*1988*
Kuwait	...	...	–	–	–	–
Malta	...	...	–	–	–	–
Oman	...	...	–	–	–	–
Qatar	...	...	–	–	–	–
Saudi Arabia	...	...	–	...	–	...
Turkey	...	...	...	17.7	...	0.71
U. Arab Emirates	...	...	–	–	–	–
Yemen	...	...	–	...	–	...
Yugoslavia	0.78	0.98	19.7	17.2	0.81	0.68

Note
For the definitions of indicators, see footnotes on Table 2.2.

Source: UNIDO Industrial Statistics Database; United Nations Statistical Office; estimates made by UNIDO.

Table A.3   Spearman rank correlations[a] relating income and market size to ratios for production, consumption and trade, 1989[b]

Ratio	GDP per capita[c]	Population
Production/consumption	0.2012* (150)	0.6945* (164)
Imports/consumption	-0.0325 (150)	-0.7309* (163)
Exports/production	0.5073* (103)	-0.1689 (108)

Notes
[a] Each of the three ratios was also regressed on per capita GDP and population. The results were qualitatively equivalent to the rank correlations shown here, but the goodness of fit was very low and coefficient estimates were highly sensitive to the choice of a particular year.
[b] Figures in parentheses refer to the numbers of countries included. Asterisks denote statistical significance at the one per cent level.
[c] In current US dollars.

Source: See Table A.1.

Table A.4   Foreign subsidiaries and foreign research facilities of selected companies[a] in industrialized countries, 1987–9

Country	Merck & Co (USA)	Hoechst (FRG)	Glaxo (UK)	Ciba Geigy (SWI)	Bayer (FRG)	Takeda (JAP)	Sandoz (SWI)	SmithKline Beecham (UK)
Australia	S	S	S	S+R	S	–	S	S
Austria	–	S	S	S	S	–	S+R	S
Belgium	S	S	S	S	S	–	S	S
Canada	S+R	S	S+R	S+R	S	–	S+R	S
Denmark	–	S	S	S	–	–	S	S
Finland	–	–	S	S	–	–	S	–
France	S+R	S	S+R	S+R	S	S	S+R	S
Germany, Federal Republic of	S	x	S+R	S+R	x	S+R	S+R	S
Greece	–	S	S	S	–	–	S	S
Ireland	S	–	S	S	–	–	S	S
Italy	S	S	S+R	S+R	S	S	S+R	S
Japan	S+R	S+R	R	S+R	S	x	S+R	S
Netherlands	S	S	S	S+R	S	–	S+R	S
New Zealand	–	–	S	S	–	–	S	S
Norway	–	–	S	S	–	–	S	–

240

Country	Merck & Co (USA)	Hoechst (FRG)	Glaxo (UK)	Ciba Geigy (SWI)	Bayer (FRG)	Takeda (JAP)	Sandoz (SWI)	SmithKline Beecham (UK)
Portugal	S	S	S	S	–	–	S	S
Spain	S+R	S	S+R	S	S	–	S+R	S
Sweden	–	S	S	S	–	–	S+R	–
Switzerland	–	–	S+R	x	–	–	x	S
United Kingdom	S+R	S	x	S+R	S	–	S+R	x
United States	x	S+R	S+R	S+R	S	S+R	S+R	S
Others	–	S	S	S	S	–	S	S

Country	Roche (SWI)	J&J (USA)	Boehringer Ingelheim (FRG)	ICI (UK)	Sankyo (JAP)	Schering AG (FRG)	Fujisawa (JAP)	Shionogi (JAP)
Australia	S+R	S+R	–	S	–	S	–	–
Austria	S	S	S+R	–	–	S	–	–
Belgium	S	S+R	–	–	–	S	–	–
Canada	S	S+R	S	S	–	–	–	–
Denmark	S	S	–	–	–	S	–	–
Finland	S	–	–	–	–	–	–	–
France	S+R	S	S	S	–	S	–	–
Germany, Federal Republic of	S	S+R	x	S	S	x	S	S
Greece	R	S	S	–	–	–	–	–
Ireland	S	S	S	–	–	–	–	–
Italy	S+R	S	S+R	S	–	S	x	x
Japan	S+R	S	S+R	S	x	S	S	–
Netherlands	S	S	–	S	–	S	–	–
New Zealand	S	S	–	S	–	S	–	–
Norway	S	–	–	–	–	–	–	–
Portugal	S	S	S	–	–	–	–	–
Spain	S	S	S	S	–	S	–	–
Sweden	S	S	–	–	–	–	–	–
Switzerland	x	S+R	–	–	S	S	–	–

Country	Roche (SWI)	J&J (USA)	Boehringer Ingelheim (FRG)	ICI (UK)	Sankyo (JAP)	Schering AG (FRG)	Fujisawa (JAP)	Shionogi (JAP)
United Kingdom	S+R	S+R	S	x	–	S	–	–
United States	S+R	x	S+R	S	S	S	S	S
Others	S	S	–	S	–	S	–	–

Country	Dow (USA)	Akzo (NL)	Tanabe Seiyaku (JAP)	Sumitomo (JAP)	Sanofi (FRA)	Astra (SWE)	Solvay (BEL)	Otsuka (JAP)
Australia	S+R	S	–	S	–	S	–	–
Austria	–	–	–	S	–	S	–	–
Belgium	–	S	S	–	S	S	x	–
Canada	S+R	S	–	–	–	S	–	S+R
Denmark	–	S	–	–	–	S	S	–
Finland	–	S	–	–	–	S	–	–
France	S+R	S	S	S	x	S	S+R	–
Germany, Federal Republic of	S+R	S+R	–	S	S	S	S	–
Greece	–	S	–	S	S	S	–	–
Ireland	–	S	–	–	–	S	–	–
Italy	R	S	–	–	S	S	S	–
Japan	S+R	S	x	x	S	S	S+R	x
Netherlands	R	x	–	S	S	S	S	–
New Zealand	S+R	–	–	S	–	–	–	–
Norway	S	S	–	–	–	S	–	–
Portugal	S	S	–	–	S	–	–	–
Spain	S+R	S	–	–	S	x	S	S
Sweden	–	S	–	–	–	x	–	–
Switzerland	S+R	S	S	–	S	S	–	–

Country	Dow (USA)	Akzo (NL)	Tanabe Seiyaku (JAP)	Sumitomo (JAP)	Sanofi (FRA)	Astra (SWE)	Solvay (BEL)	Otsuka (JAP)
United Kingdom	R	S	–	S	S	S	S	–
United States	x	S+R	S	S	S	S+R	S+R	S
Others	–	S	–	–	–	–	S	–

Note
[a] For some companies not all subsidiaries are reported.

Source: Companies' annual reports 1987, 1988 and 1989.

Table A.5 Foreign subsidiaries and foreign research facilities of selected companies[a] in developing countries, 1987–9

Country/area	Merck & Co (USA)	Hoechst (FRG)	Glaxo (UK)	Ciba Geigy (SWI)	Bayer (FRG)	Takeda (JAP)	Sandoz (SWI)	SmithKline Beecham (UK)
Argentina	S	S	S	S+R	S	–	S+R	–
Brazil	S	S	S	S+R	S	S	S	S
Chile	–	–	S	S	–	–	S	–
China	–	–	S	S	–	–	–	S
Colombia	–	–	S	S	–	–	S	–
Côte d'Ivoire	–	–	–	S	–	–	–	–
Egypt	–	S	S	S	–	S	S	–
Hong Kong	–	–	S	S	–	–	S	–
India	–	–	–	S	S	–	S+R	S
Indonesia	–	–	S	S	S	S	S	–
Malaysia	–	–	S	S	–	S	S	S
Mexico	S	S	S	S+R	S	S	S+R	S
Morocco	–	–	–	S	–	–	S	–
Nigeria	–	–	–	S	–	–	S	–
Pakistan	S	–	S	S	–	–	S	S
Philippines	–	–	S	S	–	S	S	–

Country	Merck & Co (USA)	Hoechst (FRG)	Glaxo (UK)	Ciba Geigy (SWI)	Bayer (FRG)	Takeda (JAP)	Sandoz (SWI)	SmithKline Beecham (UK)
Republic of Korea	–	–	–	S	–	–	S	–
Taiwan Province	–	–	S	S	–	S	S	–
Thailand	–	–	S	S	–	S	S	–
Turkey	–	S	S	S	S	–	S	–
Uruguay	–	–	S	S	–	–	S	–
Venezuela	–	–	S	S	–	–	S	S
Others	S	–	S	S	S	–	S	S

Country/area	Roche (SWI)	J&J (USA)	Boehringer Ingelheim (FRG)	ICI (UK)	Sankyo (JAP)	Schering AG (FRG)	Fujisawa (JAP)	Shionogi (JAP)
Argentina	S	S	S+R	S	–	S	–	–
Brazil	S	S+R	S	S	–	S	–	–
Chile	S	S	–	–	–	–	–	–
China	–	S	–	S	–	–	–	–
Colombia	S	S	S	–	–	S	–	–
Côte d'Ivoire	–	S	–	–	–	–	–	–
Egypt	S	S	–	–	–	–	–	–
Hong Kong	S	S	–	S	–	–	–	–
India	S	S	–	S	S	–	–	–
Indonesia	S	S	–	–	–	S	–	–
Malaysia	S	S	–	S	–	–	–	–
Mexico	S	S	S	–	–	S	–	–
Morocco	S	S	–	–	–	–	–	–
Nigeria	S	S	–	–	–	–	–	–
Pakistan	S	S	–	S	–	–	–	–
Philippines	S	S	–	–	–	–	–	–
Republic of Korea	S	S	–	–	S	–	–	S
Taiwan Province	S	S	S	–	S	–	S	S
Thailand	S	S	–	–	S	–	–	–
Turkey	S	S	–	–	–	–	–	–

Country/area	Roche (SWI)	J&J (USA)	Boehringer Ingelheim (FRG)	ICI (UK)	Sankyo (JAP)	Schering AG (FRG)	Fujisawa (JAP)	Shionogi (JAP)
Uruguay	S	S	–	–	–	–	–	–
Venezuela	S	S	–	–	–	–	–	–
Others	S	S	–	–	–	S	–	–

Country/area	Dow (USA)	Akzo (NL)	Tanabe Seiyaku (JAP)	Sumitomo (JAP)	Sanofi (FRA)	Astra (SWE)	Solvay (BEL)	Otsuka (JAP)
Argentina	S+R	S	–	–	S	S	S	–
Brazil	S+R	S	S	S	S	–	S	–
Chile	S	S	–	–	–	–	–	S
China	S	S	–	–	–	–	–	–
Colombia	S+R	S	–	–	–	–	–	–
Côte d'Ivoire	–	–	–	–	–	–	–	S
Egypt	–	S	–	–	S	–	–	–
Hong Kong	S+R	S	–	–	–	R	S	S
India	S	S	–	–	–	–	–	–
Indonesia	S	S	S	–	–	S	S	S
Malaysia	S+R	S	–	S	S	S	–	–
Mexico	S	S	–	–	S	–	S	–
Morocco	–	S	–	–	–	–	–	–
Nigeria	–	S	–	–	–	–	–	S
Pakistan	S	S	–	–	S	S	–	–
Philippines	S	–	–	–	–	–	–	S
Republic of Korea	S	S	–	–	S	S	–	S
Taiwan Province	S	S	S	S	–	–	–	S
Thailand	S	S	–	S	–	S	–	S
Turkey	–	S	–	–	–	–	–	–

Country/area	Dow (USA)	Akzo (NL)	Tanabe Seiyaku (JAP)	Sumitomo (JAP)	Sanofi (FRA)	Astra (SWE)	Solvay (BEL)	Otsuka (JAP)
Uruguay	–	–	–	–	–	–	–	–
Venezuela	S	S	–	–	–	–	–	–
Others	S	S	S	–	–	–	–	–

Note
[a] For some companies not all subsidiaries are reported.

Source: Companies' annual reports 1987, 1988 and 1989.

Table A.6   *Manufacturing and R&D costs as percentage of gross output, for selected industrialized countries, 1975 and the latest year*[a]

Country	1975				Latest year			
	Inputs	Labour	R&D	Other	Year	Inputs	Labour	R&D	Other
Australia	49.3[b]	23.8[b]	2.1[b]	24.8[b]	1985	48.6	18.6	3.0	29.8
Austria	67.9	17.5	4.4	10.2	1985	74.2	15.0	4.7	6.1
Canada	43.1	21.4	2.7	32.8	1987	39.4	16.1	2.9	41.6
Czechoslovakia	61.1	9.3	– 29.6 –		1989	72.2	9.0	– 18.8 –	
Denmark	40.3	27.5	8.6	23.6	1989	36.1	20.6	11.5	31.8
Finland	37.0	23.0	10.0	30.0	1985	36.2	19.5	11.0	33.3
Germany, Fed. Rep. of	51.8	24.7	11.5	12.0	1988	36.2	22.0	16.1	25.7
Greece	52.8	18.7	– 28.8 –		1985	71.7	16.9	– 11.4 –	
Hungary	53.3	5.4	– 41.3 –		1989	57.0	7.3	– 35.7 –	
Israel	58.8	16.1	– 25.1 –		1987	51.5	24.5	– 24.0 –	
Japan	37.3	12.4	5.7	44.6	1989	30.5	9.6	9.1	50.9
Netherlands	...	...	...	...	1987	60.6	22.7	9.4	7.3
New Zealand	72.2	20.2	2.4	5.1	1986	75.2	13.9	– 10.9 –	
Norway	64.0	23.2	7.6	5.2	1987	57.8	16.4	9.3	16.5
Portugal	53.5	23.6	0.1	22.8	1987	58.7	13.8	0.4	27.1
Spain	67.1	15.8	– 17.1 –		1987	60.3	17.5	2.9	19.3
Sweden	30.1	26.5	16.2	27.2	1987	22.7	16.6	27.0	33.7
United Kingdom	48.6	16.9	7.5	27.0	1988	35.5	16.4	14.4	33.7
United States	28.9	18.1	8.7	44.3	1988	29.5	12.9	10.8	46.9

Notes
[a] Figures are at current prices. Concepts and definitions vary from country to country. UNIDO statisticians have reconciled these differences whenever possible.
[b] Data refer to 1976.

Source: UNIDO, based on questionnaires received from national statistical offices, industrial censuses and industry sources.

Table A.7   Cost structure in Argentina, 1977 and 1982 (per cent of operating revenues)

Cost component	1977	1982
Manufacture	45.1	50.1
R&D	0.5	1.7
Marketing	17.7	20.5
Administration	20.6	11.5
Other costs	6.0	29.1
Profit and risk	10.0	-12.9

Source: For 1977, UNIDO calculations based on data cited in UNCTC (1984); for 1982, Asociación Latinoamericana de Industrias Farmacéuticas (1983).

Table A.8   Cost structure in Brazil, 1971, 1975 and 1984 (per cent of operating revenues)

Cost component	Locally-owned firms		Foreign-owned firms		All firms		
	1971	1975	1971	1975	1971	1975	1984[a]
Manufacture	29	27	36	34	35	33	35
Marketing	29	35	28	27	28	28	21
R&D	0	0	0	0	0	0	0
Administration	7	4	5	6	6	6	6
Other costs	5	3	4	4	4	4	22
Operating profit	30	31	27	28	27	29	16

Note
[a] Consolidated costs for 48 pharmaceutical firms.

Source: For 1971 and 1975, UNIDO calculations based on UNCTC (1984); for 1984, UNIDO calculations from the Asociación Latinoamericana de Industrias Farmacéuticas (1984).

Table A.9   Cost structure in Cameroon, 1989 (per cent of operating revenues)

Cost component	
Manufacture	52.0
R&D	0.6
Marketing	7.7
Administration	7.1
Other costs	26.9
Profit and risk	5.7

Source: UNIDO.

Table A.10   Cost structure in Cyprus, 1984 (per cent of operating revenues)

Cost component	
Advertising cost	0.4
Other overhead costs (incl. operating profit)	17.3
Wages of operatives	5.2
Other labour costs	2.9
Intermediate inputs	74.2

Source: Department of Statistics and Research, Ministry of Finance, Cyprus (1986).

Table A.11 Cost structure in Mexico, 1970 and 1975 (per cent of operating revenues)

Cost component	1970	1975
Manufacture	54	56
Marketing	15	6
R&D	0	0
Administration	8	13
Other costs	3	4
Operating profit	20	21

Source: UNIDO calculations based on UNCTC (1984).

Table A.12 Cost structure in Pakistan, 1981 and 1988 (per cent of operating revenues)

Cost component	1981	1988
Advertising, other selling costs	3.3	7.1
Patent, copyrights/royalties	0.5[a]	–
Administrative costs	4.0	8.6
Other overhead costs (incl. operating profit)	30.2	16.2
Employment	8.6	4.6
Intermediate inputs	53.5	63.5

Note
[a] Includes overhead costs.

Source: Federal Bureau of Statistics, Government of Pakistan (1984), *Census of Manufacturing Industries 1980–81*, Manager of Publications, Karachi and *Pakistan and Gulf Economist*, 17 March 1990.

Table A.13　　Cost structure in the Philippines, 1985 (per cent of operating revenues)

| Cost component | Cost range[a] | |
	Minimum	Maximum
Manufacture	58.6	48.4
R&D	0.0	0.5
Marketing	33.0	30.6
Administration	7.1	12.8
Other costs	1.3	7.7

Note

[a] Figures represent ranges across a heterogeneous sample of pharmaceutical firms engaged in both the local manufacture and importation of pharmaceutical preparations. All firms are similar in that they do little research.

Source: UNIDO calculations, based on Grabunada (1987).

Table A.14　　Cost structure in Singapore, 1986 (per cent of operating revenues)

Cost component	
Marketing and sales promotion	1.7
Other overhead costs	
(incl. operating profit)	71.9
Wages of operatives	1.6
Other labour costs	3.1
Raw materials	18.2
Packaging	2.4
Other inputs	
(fuel, electricity, water, transport)	1.2

Source: Singapore Economic Development Board (1987), Report on the Census of Industrial Production, 1986.

Table A.15   Cost structure in Venezuela, 1978, 1983 and 1987 (per cent of operating revenues)

Cost component	1978	1983	1987
Manufacture	47.4	44.1	55.8
Operational costs	44.1	42.7	37.4
Operating profit	8.5	13.2	5.4

Source: UNIDO calculations, based on Asociación Latinoamericana de Industrias Farmacéuticas, (1989).

Table A.16   Cost structure in Zimbabwe, 1980 and 1986 (per cent of operating revenues)

Cost component[a]	1980	1986
Advertising	4.7	3.0
Royalties	0.3	0.6
Other overhead costs (incl. operating profit)	31.4	27.3
Employment	15.5	15.7
Input materials	48.2	53.4

Note
[a] Data refer to the industries of pharmaceuticals, soap and detergents.

Source: Zimbabwe Central Statistical Office (1988).

Table A.17   Global and regional sales of pharmaceutical preparations, 1975–90 (percentage of world sales of manufacturers' price)

Region	1975	1980	1981	1982	1983	1984	1985	1986	1987	1988	1989	1990ᵃ
North America	21.6	20.2	21.9	24.2	26.2	29.3	30.3	27.6	27.0	27.2	29.4	31.1
West Europe	31.1	30.9	27.2	26.2	24.4	23.0	24.0	27.2	28.0	28.2	27.1	28.0
East Europe	14.5	12.0	11.2	11.2	10.4	10.0	10.8	9.6	8.3	7.7	8.2	5.9
Japan	8.0	12.0	14.1	14.5	15.5	14.8	14.3	17.5	18.9	19.9	19.1	18.4
Other industrial countries	1.8	1.6	1.6	1.6	1.6	1.6	1.5	1.4	1.4	1.5	1.6	1.6
Industrial regions	77.0	76.6	76.0	77.7	78.0	78.9	80.9	83.4	83.7	84.5	85.4	85.1
Latin America	7.8	8.0	8.7	6.3	6.0	5.8	5.9	5.1	4.6	4.5	4.6	4.9
Africa	2.8	2.6	2.5	2.5	2.5	2.5	2.4	2.2	1.4	1.3	1.2	1.2
Asia	6.3	7.0	7.4	8.1	7.9	7.7	7.4	6.7	6.1	6.0	6.2	6.4
China	6.0	5.7	5.3	5.4	5.6	5.1	3.4	2.6	4.2	3.7	2.6	2.4
Developing regions	23.0	23.4	24.0	22.3	22.0	21.1	19.1	16.6	16.3	15.5	14.6	14.9
World (US$ bn.)	42.9	79.0	81.5	82.7	86.7	88.1	93.0	112.9	135.8	154.1	163.5	172.7

Note
ᵃ Estimates.

Source: UNIDO data base.

References

Asociación Latinoamericana de Industrias Farmacéuticas, *Industria Farmacéutica Latinoamericana* (biannual), Buenos Aires: Asociación Latinoamericana de Industrias Farmacéuticas.

The Association of Danish Pharmaceutical Industry (1989), *Facts: Medicine and Health Care*, Copenhagen, Denmark.

Baily, M. (1972), 'Research and development costs and returns: the US pharmaceutical industry', *Journal of Political Economy*, **80**, (1).

Berlin, H. and B. Jönsson (1985), 'Market Life, Age Structure and Renewal – an Analysis of Pharmaceutical Specialities and Substances in Sweden 1960–1982', *Managerial and Decision Economics*, Vol. 6, pp. 246–56.

Bond, R. and D. Lean (1977), 'Sales, promotion, and product differentiation in two prescription drug markets', in *Staff Report to the Federal Trade Commission*, Washington DC: US Government.

Burstall, M. and I. Senior (1985), *The Community's Pharmaceutical Industry*, Brussels: Commission of the European Communities.

Burstall, M., J. Dunning and A. Lake (1981), *Multinational Enterprises, Governments and Technology: Pharmaceutical Industry*, Paris: OECD.

Chew, R., G. Smith and N. Wells (1985), *Pharmaceuticals in Seven Nations*, London: Office of Health Economics.

Clark, R. (1985), *Industrial Economics*, Oxford: Basil Blackwell.

Comanor, W.S. (1986), 'The Political Economy of the Pharmaceutical Industry', *Journal of Economic Literature*, Vol. xxiv, pp. 1178–1217.

Commonwealth Secretariat (1985), *Pharmaceutical Manufacture and Formulation in the Commonwealth*, London: Commonwealth Secretariat Publications.

Cooper, M. and A. Culyer (1973), *The Pharmaceutical Industry*, London: Economist Advisory Group and Dun and Bradstreet Ltd.

Cyprus Department of Statistics and Research (1986), *Industrial Statistics 1984*, Nicosia: Department of Statistics and Research.

Dunning, J. (1979), 'Explaining changing patterns of international production: in defence of the eclectic theory', *Oxford Bulletin of Economics and Statistics*, **41**, (4).

Farmindustria (1989), *Indicatori Farmaceutici*, Rome: Farmindustria.

Gereffi, G. (1983), *The Pharmaceutical Industry and Dependency in the Third World*, Princeton: Princeton University Press.

Grabowski, H. (1968), 'The determinants of industrial research and development: a study of the chemical, drug and petroleum industries', *Journal of Political Economy*, **76**, (2).

Grabowski, H. (1989), 'An analysis of U.S. international competitiveness in pharmaceuticals', *Managerial and Decision Economics*, special issue.

Grabowski, H. and J. Vernon (1981), 'The Determinants of Research and Development Expenditures in the Pharmaceutical Industry', in R. Helms (ed.), *Drugs and Health*.

Grabowski, H. and J. Vernon (1983), 'The impact of patent and regulatory policies on drug innovation', *Medical Marketing & Media*, Boca Raton, Fla.: CPS Communications, Inc.

Grabowski, H. and J. Vernon (1990), 'A new look at the returns and risks to pharmaceutical R&D', *Management Science*, **36**, (7).

Grabunada, N. (1987), *The Philippine Pharmaceutical Industry Fact Book*, Manila: Unit One Design Inc.

Grebner, K. von and V. Reinhard (1983), *Drug Therapy and Its Price: a Commercial Perspective of the Economic Aspects of Pharmaceutical Pricing on a National and International Level*, Montreal: Medicopea International.

Heiffer, M., D. Davidson and D. Korte (1984), 'Preclinical Testing', in W. Peters and W. Richards (eds), *Antimalarial Drugs 1*, Berlin: Springer Verlag.

IDA (International Dispensary Association), *Price Indicators* (semi-annual), Amsterdam: IDA.

ILO (International Labour Organization) (1986), *Economically Active Population 1950–2025*, Geneva: ILO.

IMF (International Monetary Fund), *International Financial Statistics* (annual), Washington, DC.

IMS (International Medical Services) *World Drug Market Manual* (annual), London: IMS World Publications Ltd.

JPMA (Japan Pharmaceutical Manufacturers Association), *Data Book 1989*, Tokyo: JPMA.

Lall, S. (1985a). 'Pharmaceuticals and the Third World Poor', in N. Wells (ed.), *Pharmaceuticals Among the Sunrise Industries*.

Lall, S. (1985b), 'Appropriate pharmaceutical policies in developing countries', *Managerial and Decision Economics*, **6**, (4).

Mansfield, E. (1968), *Industrial Research and Technological Innovation*, New York: W.W. Norton.

OECD (1975), *Transfer of Technology for Pharmaceutical Chemicals*, Paris: OECD.

OECD (1981), *Multinational Enterprises, Governments and Technology: Pharmaceutical Industry*, Paris: OECD.

Pakes, A. and M. Schankerman (1984), 'The rate of obsolescence of patents, research gestation lags, and the private rate of return to research resources', in Z. Griliches (ed.), *R and D Patents, and Productivity*, Chicago: University of Chicago Press.

Pakistan Federal Bureau of Statistics (1989), *Census of Manufacturing Industries 1980–81*, special report, Karachi: Federal Bureau of Statistics.

Pharmaprojects, (monthly), 1989, Richmond, UK: PJB Publications.

PMA, *Annual Survey Report*, Washington DC: Pharmaceutical Manufacturers' Association.

PMAG (Pharmaceutical Manufacturers' Association of the Federal Republic of Germany) *Pharma Daten* (annual), Frankfurt am Main: Pharmaceutical Manufacturers' Association.

Pradhan, S. (1983), *International Pharmaceutical Marketing*, Westport, Conn.: Quorum Books.

Redwood, H. (1987), *The Pharmaceutical Industry: Trends, Problems and Achievements*, Felixstowe, Suffolk: Oldwicks Press.

Reis-Arndt, E. (1987), *Neue Pharmazeutische Wirkstoffe 1961–1985*, pharma dialog series 95, Frankfurt: Bundesverband der Pharmazeutischen Industries.

Rigoni, R., A. Griffiths and W. Laing (1985), *Pharmaceutical Multinationals*, IRM Multinational Reports no. 3, New York: John Wiley & Sons.

Scrip, *Pharmaceutical Company League Tables* (annual), Richmond, UK: PJB Publications.

Scrip, *World Pharmaceutical News* (bi-weekly), Richmond, UK: PJB Publications.

Scrip, *World Pharmaceutical News Review Issues* (annual), Richmond, UK: PJB Publications.

Scrip, *Yearbook* (annual), Richmond, UK: PJB Publications.

Singapore Economic Development Board (1987), *Report on the Census of Industrial Production, 1986*, Singapore: Economic Development Board.

Slatter, S. (1977), *Competition and Marketing Strategies in the Pharmaceutical Industry*, London: Croom Helm.

SSCI (Swiss Society of Chemical Industries) (1988), *Drugs and the Pharmaceutical Industry in Switzerland*, Zurich: SSCI.

Statman, M. (1983), *Competition in the Pharmaceutical Industry: The Declining Profitability of Drug Innovation*, Washington DC: American Enterprise Institute.

Temin, P. (1979), 'Technology, regulation, and market structure in the modern pharmaceutical industry', *Bell Journal of Economics*, **10**, (2).

Thomas, L. (1987), 'Regulation and firm size: FDA impacts on innovation', Columbia Graduate School of Business, First Boston Working Paper Series, FB-87-24.

Tucker, D. (1984), *The World Health Market, the Future of the Pharmaceutical Industry*, London: Euromonitor Publications Ltd.

UNCTC (United Nations Centre on Transnational Corporations) (1984), *Transnational Corporations in the Pharmaceutical Industry of Developing Countries* (ST/CTC/49), New York: UNCTC.

UNDP (United Nations Development Programme) (1991), *Human Development Report 1991*, New York and Oxford: Oxford University Press.

UNIDO (United Nations Industrial Development Organization) (1983), *Prospect for production of vaccines and other immunizing agents in developing countries* (UNIDO/IS.402), Vienna: UNIDO.

UNIDO (United Nations Industrial Development Organization) (1987a), *Better utilization of medicinal plants: the phytopharmaceutical supply system in China* (UNIDO/PPD.47), Vienna: UNIDO.

UNIDO (United Nations Industrial Development Organization) (1987b), *Opportunities for the manufacture of pharmaceutical chemicals in developing countries* (UNIDO/PPD.48), Vienna: UNIDO.

WHO (1987) *'Certification scheme on the quality of pharmaceutical products moving in international commerce and text of good manufacturing practices'* (PHARM/82.4, Rev.3), Geneva: WHO.

Zimbabwe Central Statistical Office (1988), The Census of Production 1986/87, Harare: Central Statistical Office.

Index